THE VALUE ANALYSIS PROGRAM

A How-to-Guide for Physician Leaders on Starting Up a Successful Program

Terrence J. Loftus MD, MBA, FACS

The Value Analysis Program: A How-to-Guide for Physician Leaders on Starting Up a Successful Program

Copyright © 2016

Terrence J. Loftus

ALL RIGHTS RESERVED

No portion of this publication may be reproduced, stored in any electronic system, or transmitted in any form or by any means, electronic, mechanical, photocopy, recording, or otherwise, without written permission from the author. Brief quotations may be used in literary reviews.

ISBN: 978-1-365-06306-0

FOR INFORMATION CONTACT:

Terrence J. Loftus

President, Loftus Health

60 East Rio Salado Pkwy, Suite 9033

Tempe, AZ. 85181

www.LoftusHealth.com

Dedication

This book is dedicated to Sassy, Louie, Gabe, Ellie and Lila for teaching me the value of having a family.

ACKNOWLEDGEMENTS

I want to personally thank Doug Bowen and the incredible Supply Chain Services team at Banner Health.

DISCLOSURES

Dr. Loftus has been a speaker for Intuitive Surgical, Inc.

TABLE OF CONTENTS

CHAPTER 1: SEVEN PILLARS	2
CHAPTER 2: PURPOSE	10
CHAPTER 3: ENGAGEMENT	16
CHAPTER 4: COMMUNICATION	34
CHAPTER 5: INFRASTRUCTURE	48
CHAPTER 6: ACCOUNTABILITY	60
CHAPTER 7: LEADERSHIP	74
CHAPTER 8: PERFORMANCE IMPROVEMENT	86
APPENDIX: COMMITTEE CHECKLIST	96

CHAPTER 1
SEVEN PILLARS

When I introduced the Seven Pillars in my first book, *The Robotics Program: A How-to-Guide for Physician Leaders on Starting Up a Successful Program*, the topic of Value Analysis came up repeatedly throughout the book. Many of the examples, presented in that book, were experiences I had while developing a Value Analysis Program, which also applied to a Robotics Program we developed. While writing that book, it became obvious to me that I was going to need to write this book. With all the talk of healthcare transitioning from volume to value, the concept of value analysis is on the minds of healthcare leadership. Much as we discovered with robotics, many healthcare organizations are struggling with why they need these types of programs. Once they realize they need a value analysis program the next question is, "How does one go about setting up such a program." My journey to discovering the answer to both of these questions was not a typical one.

A few years ago my name came up at a meeting at our corporate office regarding a physician leadership role in Supply Chain Services. I wasn't in the room at the time, but what I heard was that there seemed to be consensus that I would be the right person for that role. I've never worked in Supply Chain, I've only taken one Supply Chain Management course in my life, and I knew next to nothing about the supply chain world in healthcare. My background was in Trauma Surgery and, at the time, I worked as a Chief Medical Officer for a large community hospital. So why me? Despite my initial doubts about the position, I accepted the job. It helped that I was trained as a surgeon, since many of the issues in supply chain were related to surgical supplies. What also helped was that I had a reputation for getting along well with people and getting things done. As it turns out, these are two important qualities to identify in a person who is going to lead and build a Value Analysis Program.

This job turned out to be a great opportunity. My title was Medical Director of Surgical Services & Clinical Resources. It allowed me to see what goes into the supply acquisition and decision making process for a large integrated delivery network with over one billion dollars in supply expenses per year. I felt like I had a front row seat to one of the greatest shows on earth. It is a much more complicated world than is apparent from an outsider's perspective. The strategic planning, tactical maneuvers, contracting, negotiating, procurement and delivery of supplies in a modern healthcare system is performed, almost seamlessly, by a small army of dedicated professionals. To give you a sense of how complicated this process is, our healthcare system's Supply Chain Services managed an inventory of over 160,000 SKU's. SKU's are stock keeping units, which is a reflection of the number of unique supplies in the inventory. The next time you walk into an average size grocery store, keep in mind stores of that size carry around 30 – 40 thousand SKU's. A large healthcare system must maintain an inventory 4 – 5 times larger than this. While it is not uncommon for a grocery store to run out of a particular stock item, it is extremely rare for a healthcare system to do so. That is how good Supply Chain Services in healthcare can be. So what does one see from the front row? As it turns out, it was more than I could have imagined compared to my days as a surgeon in practice.

Shortly after I began, I met with my administrative counterpart in Supply Chain Services. We functioned as a dyad with one clinical leader and one administrative leader. In reality, my dyad partner was the operational leader and I was a clinical advisor. It worked and this was due, in large part, to the person I was fortunate to have as a counterpart. There was so much for me to learn, and he had the patience and wisdom to both teach and mentor me in this new role. I attended strategy meetings, as well as sat at the table with some of the largest and best known medical device makers and suppliers. I was also the co-lead, along with my counterpart, for a system level strategic initiative focused on supply cost savings. In two years alone, we managed to exceed our stretch target and save the system over $75 million. During this time, we developed a Value Analysis Program that played a role in that supply cost savings. It also ushered in a new era of communication and information exchange with our medical staff. This book describes that journey from a physician's perspective, and what I came to see as the essential components needed to build a successful Value Analysis Program.

So for whom is the book written? It is for anyone who is interested in developing or improving their Value Analysis Program. According to the Association of Healthcare Value Analysis Professionals (AHVAP) value analysis is "A systematic process to review clinical products, equipment and technologies to evaluate their clinical efficacy, safety and impact on organizational resources[1]". Most commonly this process is implemented through a program administered by a hospital or

---

[1] http://www.ahvap.org

healthcare system. Oftentimes it is managed and coordinated either through or with the assistance of Supply Chain Services[2]. By the way, this book is not meant to replace the great work of AHVAP. This group is an excellent source of value analysis education and training. It is meant to compliment the efforts of AHVAP and others working in this area. What I hope to accomplish is to expand on the topic, and describe it from the point of view of a physician executive who has had a fair amount of success with program development. It is a summary of the journey I took, and hopefully a useful description of the lessons I learned along the way. It is the kind of book I wish was available to me when I started this adventure.

A Value Analysis Program is a facility or system based program, which is designed to review clinical products, equipment and technologies to evaluate their clinical efficacy, safety and impact on organizational resources. The fundamental purpose of any Value Analysis Program is to identify the most cost-effective supply solution. A Value Analysis Program is the overarching structure, whereas, a Value Analysis Committee is the formal executive arm which provides oversight in determining how that structure is integrated into the system and how it performs. Value Analysis Teams are the sub-groups that may be used by a program to obtain insight from subject matter experts. In the remainder of this book we will discuss the specific features of a successful Value Analysis Program and especially successful Value Analysis Committees. In the remainder of this chapter, we will focus on what I refer to as the Seven Pillars of successful programs. If you read my previous book, then this may seem very familiar. It is by design. It is my belief that all successful programs will feature the seven pillars in some manner. These are higher level elements that

---

[2] Other names for Supply Chain Services can be Materials Management or Supply Chain Management. Throughout this book, I will refer to it as Supply Chain Services.

successful programs build into their structure. To remember the Seven Pillars, think of the mnemonic "**SPECIAL PI**". This stands for, **S**even pillars, **P**urpose, **E**ngagement, **C**ommunication, **I**nfrastructure, **A**ccountability, **L**eadership and **P**erformance **I**mprovement. Chapters two through eight will provide a description of each element in detail. The final chapter is a checklist for how to start-up a successful program and specifically a successful Value Analysis Committee or Team. The following is a summary of the seven pillars.

THE SEVEN PILLARS

1) **PURPOSE:** There are key elements to any successful program. It begins with having a purpose. Everyone on your team must understand the "why". In chapter 2, we discuss the why in detail. In summary, it is because a formal organized approach to value analysis produces better outcomes at a lower cost compared to not having a program.

2) **ENGAGEMENT:** Engagement has become a buzz word in the healthcare industry. More specifically hospital administrators and physician leaders want "physician engagement". For many it has become a quest for the Holy Grail. The unstated belief is, "If we only had physician engagement, then we could solve all of our problems in healthcare." There is no doubt that physician engagement is important for change management in healthcare. Before we get to this place though, we need to set up our program to become engaging. Chapter 3 will discuss this in greater detail. In summary, the first step in this process is to stop doing those things that disengage people and move on to those things that create an engaging program.

3) **COMMUNICATION:** A Program must communicate with its stakeholders. It is impossible to educate and inform people without a consistent process for communicating to them. Communication must be a two-way process. A program must not only communicate to its stakeholders but its stakeholders must have a mechanism to communicate to the program's

leadership. Chapter 4 covers more specific information on how to communicate effectively. In summary, effective communication increases stakeholder support. If you want to be successful, then you must communicate effectively.

4) **INFRASTRUCTURE:** All programs must have a well-developed infrastructure. A program's infrastructure is the foundation on which the program will develop and sustain itself. This is often considered the boring work of building a program and therefore nonessential. Unfortunately, many organizations overlook this important element. In chapter 5 we will discuss what exactly infrastructure is. In summary, it is all the value-added people, process and technology that increase the likelihood of a highly functioning program.

5) **ACCOUNTABILITY:** Without accountability all programs eventually fail. Value Analysis Programs perform best when the accountability is integrated into the structure and governance of the program. We will talk about the pros and cons of different models in Chapter 6. In summary, someone or some group must be accountable for the following.
   a. Implementing the program
   b. Managing the program
   c. Establishing goals for the program
   d. Creating action plans to achieve goals
   e. Reporting results to Medical Staff and Administrative leadership

6) **LEADERSHIP:** In order to successfully navigate the complexity of our modern healthcare system, there must be an individual or dyad at the helm of the program. When we think of value analysis we tend to focus on the products. While the products are an important part of any program, programs are implemented, comprised of, managed and overseen by people. People want and expect good leadership as a part of any program. We've found that a dyad leadership structure is particularly effective. This is typically made up of a physician

leader and an administrative leader with experience in supply chain in the hospital setting. Chapter 7 will discuss how leadership will impact your program. In summary, all stakeholders will have varying expectations of you as a leader. Understanding what each stakeholder will expect from you will improve your ability to lead.

7) **PERFORMANCE IMPROVEMENT:** Performance improvement can be approached from two perspectives. How will the program demonstrate value to its stakeholders on a per project basis? For example, we did eight projects over the last year which resulted in a half million dollars in savings. The second perspective is how will the program go about improving itself over time? For example, two years ago we did eight projects, which resulted in a half million dollars in savings and last year we did fifteen projects that resulted in two million dollars in savings and decreased complications by 10%. In addition, it is very difficult, if not impossible, for a program to improve without data. Data is the backbone of performance improvement. Whenever possible, the data should focus on quality, utilization, total cost of care and patient/provider experience. In addition to data, it will also benefit any program to have access to recent and reliable product information. Chapter 8 will provide a more detailed description of this. In summary, you manage what you measure. If you want to improve, then you will always want to know, compared to what?

CHAPTER 2 PURPOSE

According to the 2015 Health Care Services Acquisition Report, "Healthcare mergers and acquisitions posted record breaking totals in 2014.[3]" The heat is on. Hospitals are feeling the strain of economic pressure to do something to address the rising cost of healthcare. Mergers and acquisitions are one way to approach this issue. More commonly, before a hospital reaches a point where they must do something big, they will typically attempt to do something on a much smaller scale and keeps them competitive at the local level. This usually means cutting cost.

You may be wondering why they always start with cutting cost and not consider generating revenue. There are a few reasons. Generating revenue requires an investment of some kind, and this means allocating money or resources for this investment. There are not a lot of traditional revenue generating ideas around that haven't already been tried, whereas cost cutting ideas are easier to generate and

---

[3] http://www.businesswire.com/news/home/20150331006369/en/Newly-Published-Report-2014-Health-Care-Services#.VRv4s-FKZsZ

implement. If you are already suffering from very low or non-existent margin, then obtaining the capital to develop revenue generating ideas is not really practical. If your hospital is running a 1% margin, then it will need to generate $100 of revenue for every dollar of margin. Put another way, one hundred dollars of revenue minus ninety-nine dollars of expenses. leaves one dollar of margin, which is a 1% margin. On the other hand, every dollar of cost savings will produce a dollar of margin. It is the same reason why non-profits like donations. Every donation ear-marked for the hospital goes straight to the margin by offsetting costs. In addition, revenue takes time to develop. Cost cutting is a much quicker and practical solution for increasing or preserving margin. So even if you have a revenue generating idea, you will still need to pay for it and that money will come from your margin. So this is why businesses in general start with cost cutting ideas when their margins are being squeezed. So what costs come into play when a hospital's margin is getting pinched?

The biggest cost driver in an average hospital is people, often measured in full time equivalents (FTE's) or in total salaries, wages and benefits (SWB). In 2012 personnel costs were 54.2% of the average hospital's operating revenue[4]. Because of this, it is usually one of the first places the administrative leadership will look to cut cost. The problem with this is hospitals tend to already run lean, and cutting people gets noticed very quickly. Someone has to take care of the patients and provide support for the facility. (You can always tell when these types of cost saving efforts have gone too far. The garbage cans run over, the food trays are late and patient transport takes forever.)

The second biggest cost driver is supplies, which is why there appears to be a sudden need for Value Analysis Programs. Supply expense will

---

[4] http://www.beckershospitalreview.com/finance/10-statistics-on-hospital-labor-costs-as-a-percentage-of-operating-revenue.html

typically run between 20-30% of a hospital's operating revenue[5]. There is a lot of untapped opportunity in the average supply chain expenses. We will discuss some ideas on where this is in the remainder of the book. Surprisingly the biggest opportunity is most likely to be found in improving clinical operations and care delivery performance. What is even more surprising is that C-suites believe this as well. A 2014 survey found that 71% of C-suite executives stated as much[6]. If that is so, then why not focus there?

The short answer is because that kind of work can be very challenging. Improving clinical operations and care delivery performance is complicated. It can be done, but requires a certain level of organizational competence, resources and financial support. It also requires a culture that is motivated to make big changes in their clinical and operational practices. The good news is, developing a highly functional value analysis program can serve as a stepping stone to becoming an organization that can effectively implement this type of change. Starting with value analysis allows an organization the time it needs to develop the skill set required for large scale organizational change of this nature. It was stated earlier that the purpose of a value analysis program is for an organization to identify the most cost-effective supply solution. A second purpose is for an organization to develop the resources, financial reserve and skill sets that will help it improve clinical operations and care delivery. We will discuss this in greater detail later in the book. For now, remember, there is a relationship between quality and cost that best describes value. All cost opportunities can impact quality and all quality

---

[5] Booz & Co. The Transformative Hospital Supply Chain, Balancing Costs with Quality. 2011. http://www.strategyand.pwc.com/media/uploads/Strategyand-Transformative-Hospital-Supply-Chain.pdf

[6] http://www.beckershospitalreview.com/hospital-management-administration/where-do-hospital-c-suiters-see-the-biggest-cost-savings.html

opportunities can impact cost. Finding the right balance is the third purpose of a value analysis program. So why a program?

Programs Matter

A few years ago I was asked to become more involved in understanding robotics from a system level. We ended up forming a system robotics program. The details of that effort were the subject of my first book[7]. What we discovered was that facilities that had a highly structured approach to robotics had much better outcomes compared to facilities that were less structured. Less structured facilities had a 46% higher complication rate compared to facilities with a highly structured Robotics Program. In addition, less structured programs had 37% higher costs compared to programs which were more highly structured. This wasn't just isolated to Robotics; we also observed similar outcomes for other service lines. One in particular was an Orthopedic Joint Replacement Program, which consistently demonstrated better outcomes compared to other facilities without a program. By observing what these programs were doing, this opened up opportunities for us at the system level. Around this time, we also began experimenting with Value Analysis.

It started out at one facility and slowly developed over years to a system program. The first team looked at the supply cost of Laparoscopic Cholecystectomy at one facility with one group of surgeons. The cost savings they achieved were relatively modest compared to the overall supply expense. Encouraged by this effort, the group expanded to include surgeons from other facilities, and myself as part of the group. I would love to say we were immediately and enormously successful, but that is not what happened. There were, to say the least, growing pains. I should know, I felt all of them

---

[7] Loftus T. (2016) The Robotics Program: A How-to-Guide for Physician Leaders on Starting Up a Successful Program. Tempe, AZ: Terrence Loftus.

along the way. If there is anything I've learned from all of this, it is this. There are many ways to do things that lead to poor performance, but only a few ways that will consistently lead to improved performance. One of my goals with this book is to show you the things you will need to do, and the things you will need to avoid in order to build a successful program.

Through trial and error, we eventually developed a program that could perform. It took years. This is probably not what you wanted to hear, but that is what happened. It was an evolutionary process. We tried many things along the way, only to discover they were not working. It starts with purpose and as you will learn there are six other pillars that will play a role in your success.

As you can see it is not all about cutting costs. Everyone's eyes will be focused on this aspect at first, but the real end game is about value. When we discuss engagement in chapter 3, you will be reminded that you will need to include quality from the very beginning or you will risk disengaging your stakeholders, especially physicians and nurses. They will be making this journey with you, and better to include them from the beginning if you are in it for the long haul. So where does the long haul take us?

For the foreseeable future, the end game is population health management. Very few organizations are anywhere near this application of value analysis. The one thing that is appreciated by all who are pursuing value analysis is there is a sequential evolution from no program to a fully integrated program within a population health management structure. What makes is so difficult at this time is our data systems are not adequately developed, or robust enough to collect and analyze end-to-end outcomes and total cost of care. We are considerably better that we were twenty years ago, but for most organizations we still have some ground to cover. In order to prepare for this, we will need to work with what we have, and to endeavor to improve value as we move toward the future.

To summarize, the purpose of a Value Analysis Program is to:

1) To identify the most cost effective supply solution.
2) To develop the resources, financial reserve and skill sets needed to improve clinical operations and care delivery.
3) To find the right balance between quality and cost that will lead to the delivery of excellent patient care, and a great work environment for those who provide that care.
4) To evolve into a Value Analysis Program that will become fully integrated into population health management.

CHAPTER 3 ENGAGEMENT

Engagement is the buzz word for the 21st century in healthcare. Nowhere is it more commonly applied than in the expression "physician engagement". As discussed previously it is the Holy Grail in healthcare. If you read my previous book, then this chapter will be redundant. Engagement is a necessary feature of all programs and does not necessarily need to be differentiated based on type of program. In this chapter we will discuss an alternative approach to engagement, compared to what you have previously encountered. First we will start with a few introductory recommendations before we do a deeper dive into this topic.

The first recommendation is to quit focusing on just physicians. Too often many of us in leadership point to a lack of physician engagement as the sole source of the problems in healthcare. The thought being, if only we had more physician engagement, then we could solve all of these problems. It is a false hope. We do need physicians to help with much of what ails our healthcare systems, but we need to keep in mind that they are not the cause and the sole source of the problems.

Healthcare is complicated and has a lot of moving parts, as they say. Physicians alone are not going to solve all of them. Instead of "physician engagement" what we need is engagement. The principals of engagement work for all people, including physicians.

The second recommendation is to recognize that physicians are already engaged. In the past when I was a busy Trauma Surgeon, there was not a day when I did not feel engaged. When I wasn't at work in the operating room, clinic, ward, trauma room or ICU, I was probably still thinking about my practice in some form. I could be talking to a colleague, reading a journal article or just dreaming of what I would do in a particularly challenging case. I felt like I was always engaged, much as I do about my current work. This was true of just about every physician I worked alongside. When someone complains and says physicians are not engaged, what they really mean is physicians are not engaged in the same things that person would like them to be engaging. So how do we get people to become more engaged in things that they are not currently engaging? That is the topic of the remainder of this chapter.

DISENGAGEMENT

A colleague mentioned something to me that seemed trivial at the time. He became one of the newest members of a system group that was looking at alternative supplies in the operating room. He seemed to have a particular passion for this topic, and I was looking forward to his participation. He lived quite a distance from where the group was meeting, so we set up a teleconference to allow him and other team members to join in the discussion. He called in once and then we never heard from him again. I ran into him sometime later and asked him why he was no longer participating. He said, "every time I call into that number, I'm put on hold for what seems like forever, so I just stopped calling." That particular group had a lot of members who were chronically late so we tended to start the meeting late. It was not uncommon for the meeting to begin as much as 10 minutes late.

When this occurred, we would put the phone on hold. One big mistake leading to another big mistake.

Before we can engage people, we need to stop doing those things that disengage them from the very things we want them to be engaged. In this case, as Chair of the committee, I allowed the meeting to routinely begin later than scheduled. Since we ended up starting later each time, it encouraged members who showed up in person to arrive later each time. For our call in line, we would announce very early in the call that we were waiting on some more people to show up for the meeting. Following this we put the phone on hold. If anyone called in after this, they were not always aware of what was happening. There are two big problems with this. The first is by starting the meeting later, I was sending a message to the people who arrived on time. It goes something like this. You (who showed up on time) are not as important as the people who are late. It is never a good thing to insult the people who are prompt and ready to engage. This is an example of behavior that is disengaging. Little did I know, it was also leading to another behavior that was disengaging the people who were calling the bridge line. Over time, the membership of that group dwindled. At the time we concluded that the cause of this was a lack of "physician engagement". As you can see, this conclusion was incorrect. We started with an engaged group, and slowly and very unconsciously disengaged them over time. The problem was disengagement or better described as behavior that is disengaging.

There are common attributes to behavior that is disengaging. It is inconsistent with people's expectations, may seem trivial at the time it occurs and appears to the receiving party as inconsiderate. If I tell people the meeting will start promptly at 5:00 PM and it doesn't, then I failed to meet their expectations for being prompt. If I reward others who are late by delaying the start of the meeting, then I am inconsiderate to those who arrived promptly. Many of us run into this scenario on occasion, so most of the time we are willing to excuse this behavior. If this happens repeatedly, then we become intolerant to this behavior. People resent this and feel the need to punish the person who is doing this to them. It is also interesting to note that it wasn't just physicians who quit showing up to the meetings. It was most of the members, physicians and non-physicians. This was equal opportunity disengagement at its worse. This is what we will refer to as Category I DBs (disengagement behaviors). If performed repeatedly, then they result in disengagement of just about everyone. Category II DBs are directed at a specific target audience.

We were at a Department of Surgery meeting and the next agenda item called for a report from one of our administrative managers. Sixty seconds into the report I could tell this was not going well. The surgeons were getting restless and the Chair was motioning to everyone to settle down. What could possibly be said in under a minute that could so enrage this group. In that short time span, the words or phrases "compliance", "mandatory", "Joint Commission" and "audits" were somehow use in the very first sentence the manager uttered. While these words may seem routine to someone from the hospital administration, they tend to derail many physicians and especially surgeons. This is not the best way to start a conversation. Whatever message the manager meant to communicate was completely lost on this group. The group immediately piled on the manager before the opening remarks were completed. This led to a thirty-minute rambling, tangential heated discussion on why we even needed the Joint Commission. We never discussed what the manager was attempting to tell us. What was that you ask? Our hand washing rates were improving! The manager came to give them good news, and ends up getting verbally abused for a poor choice of words. When she gave the same presentation a week later in our C-Suite meeting, she received applause.

Category II DBs are group and sometimes individual specific. We've all encountered a situation when we are completely turned off by another person's behavior, and the people we are with are not impacted by this at all. It is impossible to know exactly what behaviors will disengage specific groups and especially individuals. As we get to know people some things become predictable. For physicians, there are predictable Category II DBs that should be avoided whenever possible. If you can state a message without mentioning or doing a Category II DB, then that is what you must do. In the previous paragraph, the story demonstrates one of the biggest verbal blunders you can do with a physician audience. Physicians value autonomy and independence. They do not like to be reminded that others can and do control their behavior to some degree. There are buzz words that trigger that sense of loss of autonomy and independence. Words like compliance and mandatory will quickly disengage this audience, especially if the discussion begins with these words.

> **Rule #1:** If you want physician engagement, then avoid behaviors and language that are known to disengage them.

Another way to disengage a physician audience is to begin a discussion with just about anything related to financial concerns. When I was in my surgical training, we were told that we were not to think about the cost of what we were doing. We were to focus on doing what was right for the patient. It seemed like great advice at the time. Fast forward 25 years, and now it is impossible to not consider cost in just about everything we do. Physicians tend not to appreciate the contributions our colleagues in finance have made to healthcare over the years. Like it or not, we are now being exposed to the world of finance on a regular basis. Despite this, we still become disengaged whenever someone leads off a conversation or presentation on cost, and especially cost reduction. It's easy to drift back to a time when we only had to think about what was right for the patient, regardless of cost. We were trained to think this way, and we will always value patient care over finance any day of the week. In business school, the rules are different. One of the first things I learned was that all roads pass through finance. (The second thing was, there is no free lunch.) If you want to get just about anything done in a healthcare system, then you will need to learn how to navigate this road through finance. So how do you have discussions with physicians about financial issues or issues that may be interpreted as disengaging?

Lead conversations with physicians with what they love, and that is patient care and quality. Demonstrating the relationship between improving quality and some topic that I am not interested in hearing, such as cost improvement, is a way to improve the likelihood that I will listen to you. Focusing only on a topic a physician finds disengaging is going to lose them very quickly. So what about those times when you must do or say something that will inevitably trigger disengagement?

When we know, or even suspect, something we are going to do will lead to disengagement, then we need to consider mitigating behaviors. These are behaviors that mitigate, or reduce the negative effect of the disengaging behaviors. In our previous example, where I put people on hold waiting for other members to arrive, I could have gotten back on the line and reminded everyone that the meeting would be delayed for a few minutes more. In addition, I could have also take the time to remind everyone that in the future, we will not wait for late arriving members, and will begin the meeting promptly. I began this practice later in my career and also included ending meetings on time. This is a reactive type of mitigating behavior. I realized I was behaving in a manner that was disengaging the people our team was dependent on for our success. I learned from it and changed my behavior in future meetings. Getting back to those situations in which we know our behavior is going to be disengaging to our audience. This is a time for a pro-active mitigating behavior.

Supply Chain Services encountered a problem while negotiating for a particular supply that was a known Physician Preference Item (PPI). In order to negotiate a lower cost, we were either going to need to pressure the suppliers to lower their cost or reduce the number of suppliers. By reducing the number of suppliers, we would open market share and the remaining vendors would be more likely to reduce their prices. We knew that eliminating vendors would likely cause some bad feelings among some of the surgeons. Instead of making a decision solely on price, we decided to bring the surgeons in on the discussion through a value analysis team approach. We ran a usage list for the PPI in question and brought together a team of surgeons who used this product category on a regular basis. This team of subject matter experts were able to help us on a contracting strategy which led to over half a million dollars in supply savings per year. This was a win-win for the surgeons and healthcare system. Based on the success of this team, we developed a Value Analysis Program that went on to become highly successful. Not all proactive mitigating behaviors enjoy this kind of success.

Once again Supply Chain found itself in a high tension negotiation. This time the majority of suppliers for a particular type of PPI were willing to meet our price point for the product. One supplier, which happened to have a small group of fiercely loyal supporters, was the lone hold-out. Supply Chain took a proactive approach and met with each of the surgeons who supported the product. While these surgeons could understand the system's dilemma, many were unwilling to support changing vendors. We knew we were going to need to walk away from the negotiating table with this particular vendor, as the price they were demanding was off the charts compared to the fair market price range. We developed a time table for the surgeons as to when we would no longer provide that product, and a plan to transition to different products. A couple of surgeons moved some of their business to the competition, however the majority transitioned to an alternative product. Eighteen months later the vendor, after losing all of their market share in a dominant healthcare system in the the market, came back with a revised offer. Had they met the original price point, they would not have lost all of that market share, and would currently be priced slightly higher than where they ended up when they returned. While a few surgeons were not happy with the original deal, most were accepting of the process. Having a defined process for this situation is what made the difference.

Rule #2: If you want physician engagement, then mitigate the effects of unavoidable actions that are considered disengaging to physicians.

The single most important thing you can do to engage people is to start by not disengaging them. There are behaviors we do that can disengage just about anyone. Not only do we need to learn how to remember to not do them, we also need to take the extra step and learn how to hardwire this into our infrastructure. So even when we do forget, we can be reminded by one of those hardwired triggers. In the appendix of this book there is a checklist. This checklist is designed to hardwire mechanisms in your process to reduce the likelihood of disengaging behaviors.

Another thing we can do is change our behavior when we realize it is creating disengagement. It may not always be obvious to us, but when people stop showing up to your meeting, then there is usually something that is either driving them away or is more engaging to them. Another way to do this is to proactively mitigate disengaging behaviors that must be done. When we know something is going to disappoint a stakeholder group in our organization, then it is time to be proactive. If you were them, then how would you like to be treated? Usually this is by being forewarned of any changes, and being provided time and resources to adapt to the change.

ENGAGEMENT

One of the biggest falsehoods I hear repeated is that physicians and surgeons in particular are resistant to change. Nothing could be further from the truth. Yes, they will be assertive. Yes, they will question the need for change, but this does not mean they are resistant. There has been more change in the last 100 years of medicine than in its entire history. In just the last 25 years, we have experienced an unprecedented amount of change in surgery. Previously most of the abdominal cases were performed in an open fashion. Today, most are performed with a minimally invasive approach. This has led to changes in instruments, equipment, medication use, hospital utilization, etc. The list goes on. During this time, physicians were always changing and always demanding to understand why change was necessary. There is nothing wrong with that. Physicians are really much more change hardy than we like to believe. They are also more engaged than we like to think.

When it comes to change management in a hospital or healthcare system there is an enormous amount of time, energy and resources devoted to physician engagement. The underlying theme of many of these efforts is, how do we get physicians to change their behavior and get them to do what we want them to do? Physicians, as we previously discussed, value autonomy and independence. They are not going to change just because you want them to change. For this sin they are branded as "change resistant", "laggards" and "Luddites". Yet, this is how some of the most change hardy people on the planet are treated. Physicians are engaged, they are just not engaged in the things a typical hospital administrator finds engaging. So how do you achieve physician engagement? You have to know what they already find engaging. How do you learn what a physician finds engaging? Well, for starters you can ask them. When you can't ask them or they are not responding to your question, there is an alternative method. It's not perfect, but it can provide some insight into what they are thinking, and this gets you closer to what they find engaging.

We were in the early phase of developing a Robotics Program for our system and wanted to pull together a group of surgeons who did robotic assisted surgery. We called around the system and received a few names. They were all gynecologists based out of two facilities. It was obvious this was not representative. We wanted broader representation and people who were truly passionate about the technology. There is a fairly simple way to identify these people. We ran a usage list. We identified all the surgeons who performed robotic assisted surgery along with their specialty, number of cases, type of cases, years with robotics privileges and outcomes. The basic list of usage was easy to pull. The outcomes took a little longer, but we were able to now get a list of our busiest surgeons with the best outcomes. We told them we were interested in developing a system robotics team looking at improving outcomes, and asked them if they had any interest in this. Guess what? They were very interested in joining this group. Why? Because they were already engaged in the topics of improving outcomes and robotics and wanted to learn more. Imagine that, instant physician engagement.

> **Rule number 3:** *If you want physician engagement, then start with what they are already engaged in doing.*

We repeated this process multiple times when we developed a Value Analysis Program. Physicians are typically recruited for the value analysis teams. When we first started the program we took advantage of this usage list approach. We could easily see who used certain products and vendors the most. We could also see what types of cases the surgeons were performing most frequently. Surgeons love the tools they use for operating, and if there is going to be any discussion regarding them, they want to be at the table. (Many have learned the hard way, that if you are not at the table, then you are on the table.) Usage lists tell you not only what types of cases interest them, but also what types of tools interest them. When you want to include a discussion on outcomes, then find those physicians who are already getting great outcomes. They are almost always interested in improving. Why? Because that is what they are already engaged in doing.

We were implementing an enhanced recovery after surgery program for bowel surgery across our system, and trying to determine if there was any support for the program. Once again, we turned to the data. We planned on monitoring two key steps in the program. They were early ambulation and early alimentation. We ran a list of surgeons who performed these types of cases and a list of how often these two key steps were being performed by their patients. Overall these steps were being performed only 30-40% of the time on average. What was really interesting was that a majority of the surgeons were ordering early ambulation and early alimentation much more frequently than this. Evidence based medicine supported this practice. The surgeons were aware of the evidence, and thought they were doing it as best they could. Unfortunately, the actual performance and outcomes were not matching this expectation. When we implemented the program we built decision support into the order sets and created a tracking board of these patients to make the individual patients' performance more visible. Not only did the performance of these key steps improve significantly, but is was also associated with a significant improvement in complications and readmission rates[8].

> **Rule #4:** *If you want physician engagement, then help them solve a problem they are already attempting to solve.*

For ten years in a row we endured continuous price increases in a particular product category. So we ran a usage list for this product, and pulled together a group of surgeons with the greatest use. What was particularly troubling about this product was the hospitals lost money every time this product was used. The amount was substantial and was creating a risk for the service line. We were at the breaking point, and were not going to be able to support this product and the surgeons anymore to the degree we did in the past. There were a lot of alternative plans discussed, but we wanted to know what the surgeons thought. They were appalled. They were also unaware this

---

[8] Loftus T, Stelton S, Efaw BW, Bloomstone J. A system wide care pathway for enhanced recovery after bowel surgery focusing on alimentation and ambulation reduces complications and readmissions. J. Healthcare Quality. Published online Feb 20 2014. doi: 10.1111/jhq.12068

whole time how much the product cost, and that the hospitals were losing money every time it was used. With the support of the surgeons, we approached the vendors regarding a significant price decrease. Initially the vendors were not interested and wanted another price increase. They appealed to the surgeons for support. Not one surgeon would support the vendors, and told them so in no uncertain terms. We ended up with the price decrease we needed to sustain the program.

> **Rule #5:** *If you want physician engagement, then appeal to their sense of fairness and what is right for the sustainability of their their practice, their service line or even the hospital in which they practice.*

There are going to be times when you must address issues that many physicians will simply not be interested. We ran into this with supply cost in the operating room. Just about every surgeon whose supply costs were on the high side was emphatic that they had to use higher cost supplies because they were getting better outcomes. Specifically, they insisted their complication rates were lower than their lower cost peers. Physicians are scientists by training and this was a testable hypothesis, so we pulled the data. We looked at the complication rate for Laparoscopic Cholecystectomy. In addition, this data was risk-adjusted and bench-marked to Premier's select data to reveal each surgeon's actual to observed ratio for complications. (See Graph 1)

Less than one is better than expected, and greater than one is worse than expected. We than compared this to the surgeon's average supply cost and ranked them by percentile from best to worst (left to right on Graph 1). There is no correlation between supply cost and complications. Needless to say, this generated some physician engagement, but it was not enough for most surgeons to want to run out and change their supply usage.

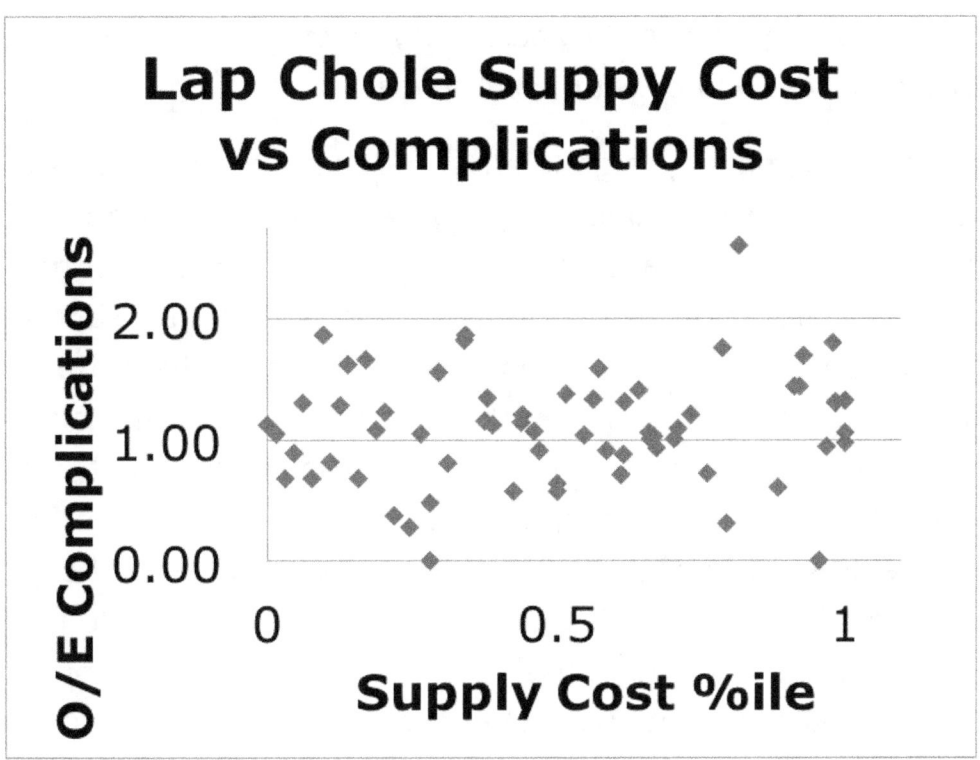

Graph 1: Comparison of average supply cost per surgeon to their observed to expected complications for laparoscopic cholecystectomy

We then suggested that one of the ways they could reduce their supply cost is to start using reprocessed equipment. There was a great deal of heated discussion regarding reprocessed supplies and the consensus among the surgeons was that reprocessed (RP) supplies were defective more frequently than original equipment manufacturer (OEM) supplies. This too was a testable hypothesis. We tracked the surgeon reported defect rates for single use energy devices. What we discovered was that the OEM defect rate was 4.9X greater than the RP defect rate[9]. Our reprocessing usage rate improved dramatically after this announcement as well as the associated supply cost savings.

---

[9] Loftus, T. A Comparison of the Defect Rate between Original Equipment Manufacturer and Reprocessed Single-use Bipolar and Ultrasound Diathermy Devices. J. Med. Devices. 2015; 9(4):044501-044501-2. doi:10.1115/1.4030858.

Rule #6: *If you want physician engagement, then appeal to their interest and passion for science.*

Informational asymmetry is generally considered a competitive advantage. This is where one company has access to information that another company cannot access. It is an important mechanism in a competitive environment. Taken outside of this context it can have unintended consequences. For years' suppliers in the healthcare industry insisted on statements in their contracts with hospitals and healthcare systems, which prevented them from discussing the prices of their products with their competitors. Understandable from a business perspective, however one of the consequences of these types of agreements is hospitals felt they were therefore prevented from having discussions with their physicians regarding the costs of supplies. This is simply not true. Hospitals can have these discussions with physicians. Unfortunately, this belief led to decades of physicians being prevented from understanding the cost of the supplies they used. Physicians have some understanding of the quality of the products they use. When the costs are hidden from them, it is almost impossible for them to understand the value of them.

While costs in healthcare are complicated, they are not too complicated to be understood by physicians. It just takes time and an interest to learn. We provided a group of General Surgeons with their average cost per case for Laparoscopic Cholecystectomy. This alone, was insufficient to impact any change in their average cost. In a blinded fashion, we then showed them their average cost per case compared to their peers. Something in them changed. They were now very interested in cost. At one point, surgeons were now debating over who had the lowest cost per case. When they found out who it was, they wanted to see what it was that surgeon was doing. Their concern was how could this one surgeon possibly be beating them on average cost and still be getting good outcomes. For Laparoscopic Cholecystectomy most of the difference in cost comes down to a few instruments and several preference card practices, which we will discuss in a later chapter. Over the next several months the group reconciled the differences on their supply usage. They ultimately developed one single, low cost, preference card for the group. The group now sees themselves as the high value (high quality, low cost) alternative and markets themselves to payers as such.

> **Rule #7:** *If you want physician engagement, then appeal to their competitive spirit.*

This particular success story happened for another reason as well. The group was motivated by their competitive spirit. They not only wanted to be the best as individuals, but along the way they discovered a group identity that now wanted to also be known as the best group. Motivation only gets you part of the way. You can have all the motivation in the world, but if you do not have the ability to change, then change will not occur. So where did they get the ability to change. It came from themselves. They could see who among them had the low supply cost and best outcomes, and were able to model the group on the most cost-effective supply solution.

> **Rule #8:** *If you want physician engagement, then empower them to create their own solutions.*

Just as I was leaving work to go on vacation, I received a call from a very irate surgeon. He was upset and he needed to talk to someone from the administration. I was the Chief Medical Officer, so I was used to these types of calls. This one was a little different. He spent the first half of our hour long conversation dropping just about every imaginable f-bomb you can hear. He was so loud and upset, I had to hold the phone away from my ear. During this time, I was occasionally scribbling notes on a post-it note. The last half hour of our conversation was more amicable. We agreed to talk when I returned the following week. When I returned a week later there was a stack of letters, memos and reports I needed to review along with the other duties of my role. It was at least a week later when I found a yellow post-it note with a couple of ideas written on it. When I first read them, I thought, "these are some great ideas, why didn't' I think of them sooner". Just then it hit me. These were the scribbled notes I wrote when listening to the irate surgeon during the prior week. Buried beneath his anger, disappointment, negativity and multiple f-bombs were a couple of great ideas. Even though he was extremely upset, his ideas were what he was really trying to communicate. His passion was obscuring the real contribution. There are going to be times when you have to filter what you are experiencing when confronted with very passionate and angry staff. Beneath that passion you could very well discover some great ideas and engaged physicians.

> **Rule #9:** If you want physician engagement, then listen to them when they are upset, because they are already maximally engaged and may be harboring ideas worth considering.

In the previous section on disengagement, I mentioned the example of the manager who approached the Department of Surgery with disengaging verbiage such as "compliance" and "mandatory. What if the manager approached the group differently? For example, what if she could have stated it this way. "Thank you for allowing me to present an update from the quality department. We have some great news to report. As you are all aware there is a relationship between lower hospital acquired infections and hand washing. We studied our hand washing rates and we're happy to report there is a significant increase in hand washing for the facility. We think it has to do with the addition of new gel dispensers available outside patient rooms. We appreciate that so many of you are using these when you 'gel in' and 'gel out' as you round on your patients. This should go a long way to improving our infection rate." See the difference. In the first description the disengaging verbiage dominates the presentation. In the second description, the engaging verbiage dominates the presentation. Focusing on the positive and including language consistent with the values of the audience is more likely to have a favorable impression. In addition, the manager would have been able to describe the specific behavior needed to continue to improve the rates.

>**Rule #10:** If you want physician engagement, then communicate with words that positively reflects the things they value.

The single most important thing you can do to engage people is to listen to them and understand what they are already engaged in thinking and doing. The next most important thing you can do is to appeal to what they value. The final thing you can do is to break down the silos that create asymmetric information. This can empower them to create their own solutions. Because the staff on the frontlines have first-hand knowledge of the problems, they are more likely to implement a more cost-effective and sustainable solution.

One last word of advice regarding consistency. Good habits are the result of being consistent. By developing good habits on how you structure your program, and approach your stakeholders, you will appear consistent to them. By being consistent, you will build credibility and trust. People want to follow leaders they know they can trust, so, as a rule, if you want physician engagement, then be consistent.

CHAPTER 4
COMMUNICATION

The language used in the world of value analysis has its roots in the language of Supply Chain Services. In the first part of this chapter we will review some of the more important concepts you will need to understand in order to communicate effectively in the value analysis world. The second part of this chapter will focus on who you will be communicating, and ideas on effective ways to do this.

Learning the Language

It was my first rodeo in the Supply Chain Services world and my job as Medical Director, as I saw it, was to hold the line on quality and patient safety. I still think that is a fundamental role of anyone who assumes this role, but my first time meeting at the negotiating table with a Fortune 500 medical device manufacturer taught me something else. The vendor's team was doing a presentation and I spent the better part of the time asking questions regarding whether there was any evidence demonstrating better clinical outcomes with their products. There was a break in the meeting and one of the senior level negotiators from their team asked to speak to me in private.

He said, "Dr. Loftus, I appreciate your questions on quality and outcomes but I have to be honest with you. We already believe we have a good product. When we come to these meetings we are looking for something else from your team."

"What's that," I asked.

"Well, if my team does not hear the words 'volume' or 'market share' come out of your mouth in the first thirty seconds, we tend to stop listening to you," he said. "it's not that we aren't sensitive to your concerns, it's just that in order for us to reach the price point your team is requesting, I need to get permission from our pricing council. They will not listen to a word I say unless I use those words. They will specifically want to hear how your company will increase the volume or market share of our products in order to justify a price reduction."

That just blew me away. I was speaking a language that was disengaging to the vendor. I spoke to my counterpart in Supply Chain Services about this conversation, and he confirmed that is just the way they view product negotiations. It comes down to volume and market share for them. Quality and outcomes are just not a regular part of their conversations, unless of course, they are speaking to a physician away from the negotiating table. In that case, they never discuss volume and market share, and focus on what they consider are advantages of their product. Most often this will include features that a physician will appreciate, such as how the instrument feels and performs during procedures. Due to the lack of Level I evidence for many medical devices, the topic of clinical outcomes rarely, if ever, is discussed with physicians. They have learned many years ago to stay away from any discussion suggestive of finance (volume and market share). This, of course, would be a disengaging behavior to a physician (DB Type 2) from their perspective, and should be avoided.

There are many important lessons to learn from this story. I don't want you thinking physicians should avoid conversations regarding quality and patient safety with medical device manufacturers. Keep in mind, that was one of the reasons for me to be involved in the negotiations. Vendors are not used to having physicians involved in these conversations, so it is new for them, just as it is new for physicians to be sitting at the table when these conversations are occurring. What is important to learn is to understand what has, up until recently, been driving decisions regarding product procurement.

I also learned a very valuable lesson that could be used in future negotiations. Vendors are very motivated to increase the volume and market share of their products. There is also a very interesting psychological effect that can be used in this situation. It is loss aversion. Loss aversion is the observation that people prefer avoiding losses to acquiring gains[10]. Simply put, vendors will do almost anything to avoid losing volume or market share and that includes price point concessions and other types of compromises. In one negotiation, we enlisted a group of surgeons to side with the system. Two of the four vendors, with products in question, were willing to walk away from the table because we were not conceding to a price increase. We made it clear that not only were we going to let them walk away and drop their contracts, but that we would hold them to it for the terms of the contracts we had with the other vendors who were willing to work with us. That meant they would be off contract for two years. That represented a lot of volume and market share to them and the pain of losing that much was unbearable. They agreed to the price reduction we requested, which was fair market value for their products. That would have been a very difficult negotiation without the support of the surgeons.

There are a few things to learn about price points. First off, a price point just refers to a negotiated price between a supplier and a customer. While it is not possible to know what prices a vendor is charging the competition, there is a way to learn how to get a reasonably fair price. If you have ever used Kelly Blue Book or True

---

[10] https://en.wikipedia.org/wiki/Loss_aversion

Car websites, then you will understand the concept. If you go to one of those sites, they can give you an estimate of what a vehicle is worth in your market. It will show a range of prices and a market average. A similar methodology can be used for medical supplies. Hospitals can get a sense of what is a fair price for products compared to others in the market. If a vendor is requesting a price point that is off the charts with regards to the market, then it is okay to call them on it. Most are aware of this information and are usually willing to start negotiations within an acceptable range of price points. The art of the negotiation is to determine what is an acceptable price point for both sides. So in the previous example, we knew that two of the vendors were not negotiating in good faith. Their requests were off the charts. We knew it and they knew it. We gave them two days to come to their senses. We were offering them a fair market price and coupled with the risk of losing volume and market share, they finally came back to the table.

There is another important lesson to learn about pricing. The price listed in your company's computer may not be the real price. This is how that works, and why you may never trust another paper in the literature that compares supply costs. There is more on the table when price points are being discussed. For example, if the supplier and the hospital agree that a price point for a particular product is $100, then most people will assume that is how much the product will cost the hospital. This is not always true, and here is why that is. These agreements can also include language that provides incentives for volume. For example, if the hospital purchases the product in bulk, then the supplier may apply a 5% savings. If they order the product one at a time, then the price is $100. If they order the product 100 at a time, then the real price is $95. In addition, they can also provide incentives in the form of rebates. If the hospital purchases 1000 of these products over the next year, the supplier may provide a 10% rebate. The real price is $90. Another way the price can be offset is if the product is bundled with other products the supplier offers. The product's price remains $100 without bulk purchasing or rebates, but is actually lower because it is bundled with other products that includes great price points for the other products. As a bundle, the $100 product may now really be closer to $85. The listed price remains $100, but this may be the better deal for the hospital. Keep in mind, if you publish a paper in the literature using administrative data, then you may really be reporting results that do not reflect the total true negotiated cost.

So why would a hospital agree to deals like this? The answer is, to obtain better pricing. The better question is; why would suppliers be willing to go to such lengths to protect a price point? Remember the True Car example? Where does this type of information come from? If your hospital is working with a Group Purchasing Organization (GPO), then it is probably from the GPO. A GPO is "an entity that is created to leverage the purchasing power of a group of businesses to obtain discounts from vendors based on the collective buying power of the GPO members[11]". The GPOs obtain their information from the administrative databases of the members. So if the list price remains at $100, then that is what the other members will see when they query the data from the GPO. This is a simplification of how hospitals obtain information on fair pricing, but it is useful way to understand the fundamental process. The supplier's goal is make sure that other hospitals are not aware of your hospital's real pricing, when you are getting a better deal than the other hospitals are getting.

Returning to our True Car analogy, let's look at pricing from a strategic position. Once you see the results of the range of prices being offered, you will need to develop a strategy regarding which price is the best price for you. If you are very selective about which specific features you want in a car, then you will probably need to pay at the higher end of the range. If you are less picky about your selection, then you will be okay starting at a lower initial offer. A similar thing happens with

---

[11] https://en.wikipedia.org/wiki/Group_purchasing_organization

hospitals. If a healthcare system is very large and controls a large market share, then it will make an offer toward the lower price range. If it is a small single hospital in a large urban area with small market share, then its negotiating position can be diminished by this, and will most likely be paying a much higher price.

Hospitals can also create market share. This can be done by eliminating competition for one or more of the suppliers. If you currently have multiple vendors for a product category, a hospital can open up market share by signing contracts with a limited number of suppliers. The ultimate expression of this is to "sole source" your contract. This means just one supplier is providing for this product category. This is a great way to get the best price. The downside of it is the suppliers who were eliminated may have products that are preferred by physicians. One way around this is to "dual source". Two vendors that make up the majority of the physician preference items assume the contract. This reduces the number of physicians impacted by a supply decision and the hospital obtains better, but not necessarily, the best price.

If a product category has multiple vendors, and they are all preferred by multiple physicians, then one way to approach this is to do an "all-play" strategy. Returning to our True Car analogy, a fair (median) price point is selected and all the suppliers are asked to participate at that price point. If a supplier is not willing to accept that price point, then they are off contract for the term of the contract (usually 2 years). This way it is the vendor who makes the decision to lose volume and market share, which we know they will generally hesitate to do.

One last thing before we move on to a general discussion of communication. It is important to understand that just about everything in Supply Chain Services is ruled by the contract calendar. What this refers to is that just about every product in the hospital's inventory will need to be on some type of contract. It is the contract with the supplier that not only determines the price point but also any specific provisions that will allow for bulk buys, rebates, bundling, volume or market share requirements, in addition to the conditions for contract termination, and the exact date the contact ends. Contracts typically range from one to three years in length and average around 2 years. Hospitals will negotiate a longer contract when they get better pricing and a shorter contract when they believe the pricing is not optimal or there are concerns that a new product is in the pipeline and they do not want to miss the opportunity to get out of a contract quicker. Contracts cost money to negotiate and review, so longer contracts are favored by hospitals and suppliers for this reason. (In our system, our contracting team could easily be reviewing a dozen or more a week.) Therefore, both groups are incentivized to not deviate from the contract calendar. It is almost always better to stick with this calendar. This allows Supply Chain Services to plan ahead of time. When you are developing a Value Analysis Program this will be a key variable in your planning as well.

Communicating with Stakeholders

A thorough understanding of your stakeholders is the beginning of any effective communication plan. Leave one of them off the list of those with whom you will be communicating, and it could create some real problems for the success of your program. We're going to start with the most obvious group first, and that is the members of your Value Analysis Committee. They are not just members of your committee, they are also your program's ambassadors to the hospital and community. If this group does not understand, and cannot articulate what the program is doing, then other stakeholder groups will be receiving conflicting information.

Value Analysis Committee Communication

There are two basic documents used for communicating with your members. They are an agenda and meeting minutes. Do not neglect the importance of these documents. The agenda is a forward looking document. It describes what you will be addressing in the future. The meeting minutes are a backward looking document. It describes what we discussed and found important enough to put in writing. One looks at what we would like to accomplish as a team, and the other looks at what we were able to accomplish as a team.

There are two bad habits that committees can get into with these documents. The first is regarding the agenda. There is a tendency for committee leaders to issue the same agenda for every meeting. This sends the message that this next meeting will be no different than the last meeting. This is a disengaging message to your membership. The agenda must change if the program is to change. Ask your members for agenda ideas. Bring in a speaker to talk to your committee. There are going to be certain agenda items, like your dashboard, that will be recurring items. The point is to mix it up and make it interesting, even engaging, for your membership. How do you do that? Provide new information for the group. There are several excellent sources of information available, such as Hayes[12], ECRI Institute[13] and AHVAP[14]. This could also be a recent journal article, information from a meeting on a new product, a video from YouTube demonstrating the advantages of a product. Make it interesting and they will want to be there.

The other bad habit has to do with the minutes. They are either too short and provide no useful information, or too long and detailed. Both types tend not to get read, although always approved. You want your minutes to reflect the discussion and especially any conclusions or action items for the team. Years from now, you should be able to look at any big decision the committee made, and understand why the committee made that decision. This is a very effective way to communicate to future leaders and members of your committee. Minutes aren't just for the current leaders and membership; they are

---

[12] http://www.hayesinc.com/hayes/
[13] https://www.ecri.org/
[14] http://www.ahvap.org

for all groups who want to understand why things happened the way they did. What makes perfect sense today, may not be relevant in five years.

Members should have the ability to contact one another on a regular basis. Create a membership list with contact information and distribute the list to all members (with their permission). There are going to be times when members will want to have an offline meeting to discuss some issue that they did not feel comfortable discussing in an open meeting. An offline meeting is a meeting that occurs between two or more members outside of the scheduled meeting. Minutes are not recorded for these types of meetings. These types of meetings are for members to clarify and work out any issues that were not, or are not going to be discussed in the open meeting.

One final note on Value Analysis Committee communication and that is where to store all official documents for the committee. Ideally this should be a cloud based repository (Dropbox, Google documents, etc.) or SharePoint. Committee members work all hours of the day and will need to be able to access this source of information. While it is important to email or otherwise distribute meeting materials, it is also essential for members to be able to review this information from anywhere and anytime.

Medical Staff Communication

Probably one of the most underappreciated teams in any hospital is the Medical Staff Services team. Much of what you read written in this book regarding infrastructure and communication was taught to me over the years by the people who ran the Medical Staff Services offices in the healthcare organizations where I worked. If your Value Analysis Committee needs to communicate to the Medical Staff, and by now you should know you will need to communicate to them, then you will need to bring the Medical Staff Service's office onboard early in the process. Regardless of whether you use a Centralized Model or a

Decentralized Model (more on this in the chapter on Accountability), it will be to your benefit to involve this team.

Medical Staff Services have mastered the art of organizing, communicating and coordinating the affairs of physicians and providers for decades. It has been referred to as "herding cats", but this is not accurate. It just seems like herding cats when you're not very good at it. Because of the skill set they bring to the table this team can show you the way.

Here is one way they can do this. Early in the develop of our system Value Analysis Program, we needed a way to communicate to physicians. I contacted our Director's Group for Medical Staff Services and requested the email contact information for all of our Medical Staff. They refused. 'Why", I asked. "Because that is confidential information", they responded. Good point, I forgot about that. They also pointed out that the list changes every day and it would be impossible for me to update it all the time. Good point, again. They reminded me that some of the physicians do not check their email or spam all the email coming from the hospital. Didn't know that, but also another good point. What do they do in that situation? It depends. Sometimes they send an email to the physician's secretary to print it out and hand it to them. Sometimes they catch the physician in the lunch room or rounding on patients and talk to them. The bottom line is they have learned how to contact every physician on staff, when they really need to contact them. We then worked out an arrangement where I would forward the Directors of Medical Staff Services a newsletter that summarized the committee's finding. They would then forward it to the appropriate physicians and Medical Staff meetings. Sometimes it doesn't always work as planned, but it is safe to say, it was by far, one of the most consistent ways I found to communicate to large groups of physicians.

While this method is consistent the real question is, was it effective? Physicians are busy people, and let's face it, their time comes at a premium. When you have their attention, maximize the use of that time. Wasting it, is another form of disengaging behavior. At the end of this chapter is an example of a newsletter we did for supply chain services. Note a few things. All the monthly news fits on one page. Each section is clearly marked so a physician can look at this with one glance, and in a few seconds see if there is anything of interest to them. The most important words and sentences are bolded so a physician can quickly read the highlights of the section to see if the information applies to their practice. If there is more information, then a memo is provided along with the newsletter and is available to the physicians who want to read additional information. There are links in the newsletter so if a physician is interested in additional information, then they can simply click on the link. This is when a SharePoint is especially useful because you can store and link the document in one place. This newsletter also includes an email link to the leadership in case a physician has any questions or comments. Send a few of these out and notice the response. You will quickly find out if you are being effective or not. Physicians tend not to hold back when they really want to get a point across.

If only newsletters and memos were all it took to communicate to your Medical Staff. Nothing does it better than a face-to-face conversation. There are going to be times when you must get up in front of a group of your peers and explain why you are doing what you are doing. They may not agree with you, but if you are not willing to face them, then they will lose trust in you. When it comes to the relationships you have with your Medical Staff, trust is the cornerstone. The CEO of a company where I worked used to remind us of an old saying. Visibility leads to credibility, and credibility leads to trust. If you want to be trusted, then you must be visible. If your Medical Staff are complaining about communication, then it is time to get visible.

C-Suite and Senior Management Communication

If the Administrative Lead on your committee is a member of the C-Suite or reports to Senior Management, then this will be very easy. If they are not, then you will need to make sure that what is happening on your committee is being communicated to them. The C-Suite is made up of the senior management team for your facility. They will be involved all of the major decisions regarding operations and capital purchasing. They will also need to be informed regarding any significant changes in supply purchases. This will be especially important for physician preference items (PPIs). Why is that? Because if a PPI is changed and a physician is not happy with this decision, they will often make their feelings known to senior management. If your C-suite is informed of why this decision was made, then they can anticipate this response and help you proactively manage it with the Medical Staff. While they will enjoy hearing about what a great job you are doing to improve clinical outcomes, they will be especially interested to hear about the improvements in utilization and cost reduction. They, after all, want to know what the return on investment is. This is also a great opportunity to let them know what resources you need. Your Administrative Lead should be running point on this. If you are not able to improve your performance because you lack adequate resources, then this team needs to be aware of this situation. One last word of advice, whatever you are communicating to your Medical Staff, it must me the same to your C-Suite. Believe it or not, they talk to each other. Well, at least in most organizations. So don't forget to include your C-Suite on the Newsletter or any memo distribution list.

Nursing & Ancillary Staff Communication

This tends to be easier to do than physician communication. Just because it may be easier does not mean it can be neglected. Typically, the Chief Nursing Officer and this person's direct reports are your best avenue for communicating. Coordinating with this person early and regularly is important. You can use a newsletter or any other approved facility communication tool. Be open to feedback, particularly when it comes from the operating room and Cath Lab. If you do not have your OR or Cath Lab Directors on your committee, then this is a good time to include them. This person will know the key players and how other operations will impact your program.

Supply Chain Service's Newsletter

What's New in November, 2014

Orthobiologics: DBM

- Earlier this year an Orthobiologics **Value Analysis Team composed of Surgeons** met to discuss the benefits and cost-effectiveness of multiple products in this supply category. Based on the **recommendation** of this team, effective immediately, **The system will no longer purchase or contract for the following Demineralized Bone Matrix (DBM) products:**
    - Stryker AlloFuse Gel, AlloFuse Putty
    - Stryker AlloFuse Plus Paste, AlloFuse Plus Putty
    - Stryker AlloMatrix Putty
    - Stryker Allograft Wedges
    - Stryker DBM Gel, Putty & Putty Plus
    - Integra Dynagraft products
- **All other contracts for DBM products remain in effect.** We do have existing contracts with Bacterin, Medtronic, MTF, and others for DBM products and will continue to support this product category.
- **See attached memo.**

Irrigation Fluid Shortage

- **Conservation methods are working.** At our current rate of use we are on track to meet the usage needs into the foreseeable future. Since this shortage is anticipated to last into the first quarter of 2015, **we are asking everyone to continue to conserve irrigation fluid.** The effort on the part of the physicians, staff and Supply Chain has made all the difference. **Thank you for your assistance!**

Supply Chain Services Link

Additional Information
- Click on this link

Supply Savings Strategic Initiative

- The **Overall Supply Savings Strategic Initiative** results, as of Nov 3, 2014, are in. **We exceeded our stretch target** for the system and are at $28,632,916 for the year. Peri-op is 106% of the way at $7.54M and Cath Lab is 126% of the way at $6.09 M to their target. **Congratulations** to the facility and discipline teams for their outstanding performance!

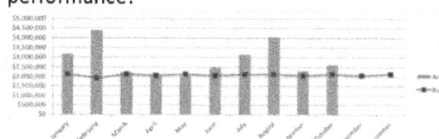

Supply Chain Education & CME

The **American Association for Physician Leadership** (formerly ACPE) is offering a course titled, **"Physician Leader's Role in Supply Chain Performance"** (CME = 14hrs) at the Fall Institute in Scottsdale, AZ (Fairmont Scottsdale Princess) on Nov 16–17. Contact AAPL for more info.

Safe Surgery

The results of the system's Safe Surgery Program were published online in the **Journal of the American College of Surgeons.** The title is **"Implementing a Standardized Safe Surgery Program Reduces Serious Reportable Events".** There was a **52% decrease in the SRE rate.**

Questions or Feedback

Contact: Terry Loftus, MD, FACS
TerryLoftusMD@gmail.com
Medical Director
Surgical Services & Clinical Resources

CHAPTER 5
INFRASTRUCTURE

Infrastructure is the foundation on which any program will be built. There are elements of infrastructure which will be foundational for all types of programs and elements that are specific to a Value Analysis Program. We will cover both in this chapter. All successful programs have a well-developed infrastructure. Unsuccessful programs and poorly performing programs, all too commonly, can be found to be lacking in basic infrastructure. So what exactly do we mean when we say infrastructure.

INFRASTRUCTURE DEFINED

Infrastructure refers to all the people, process, technology and cultural elements that are required for the successful implementation, management and sustainability of a program. The specific elements are categorized according to people, process, technology and culture.

PEOPLE

People are the personnel who are essential for the smooth operation of a Value Analysis Program. They are characterized by the roles and responsibilities they have as members of the Value Analysis Program.

The following roles and responsibilities are considered fundamental to any successful program.

1) **Physician Lead:** This individual provides direct oversight of the Value Analysis Program. This role can be developed as a sole leadership role or supported by a co-lead in a dyad model. Most commonly a dyad model in a hospital setting would be with a Supply Chain Services lead. This leadership role functions as the Chair of the Value Analysis Program and any Value Analysis Committee or Teams on which they serve for the Hospital or Healthcare System. They are responsible for setting the goals of the program, developing an agenda for meetings, running the meetings, developing the program/meeting membership, proposing the purpose of the program, develops goals for the program and is the primary person responsible for communicating any information regarding the program to the Medical Staff and leadership. In addition, this person is also the individual responsible for receiving and responding to feedback from the Medical Staff and leadership. The Physician lead, along with the Supply Chain Services lead, provides oversight for the Value Analysis Program's Performance improvement.

2) **Supply Chain Services Lead:** This individual, when functioning as a dyad co-lead, has the same responsibilities as the Physician lead. The Supply Chain Services lead is the person responsible for communicating any information regarding the program to the Supply Chain Services, Nursing and Ancillary staff. In addition, this person is also the individual responsible for receiving and responding to feedback from this Staff. This person should be a senior level manager in Supply Chain Services and should have direct reporting responsibilities with a facilities C-Suite (Decentralized Model) or Senior Management of the Healthcare System (Centralized Model). One of the main functions of the Supply Chain Services lead is to identify and secure resources necessary for the smooth functioning and sustainability of the

program. The Supply Chain Services lead, along with the Physician lead provides oversight for the Value Analysis Program's Performance Improvement.

3) **Nursing Lead:** This individual can be paired with a Physician lead in a dyad model to co-lead the program. If so, then the Nursing lead's responsibilities are similar to the Supply Chain Service's co-lead responsibilities. The Nursing lead's primary function is to be the liaison between the Nursing Staff and the Value Analysis Program. This individual is responsible for communicating any information regarding the program to these staff. This person should be a senior level manager who has direct access to Nursing leadership. Ideally this person should have operational and management oversight of the areas impacted by decisions made by the Value Analysis Program.

4) **Contracting Lead:** This person either provides managerial oversight to the contracting process or is actively involved in the contracting process for Supply Chain Services. The Contracting Lead's primary function is to be the liaison between the contracting process and the Value Analysis Program. This person will provide insight into the nuances of contracting as well as a historical perspective on the hospital's relationships with the various vendors. This person will maintain the contracting calendar for the group as well as support any research conducted in the evaluation of supplies being considered for the organization.

5) **Administrative Assistant:** This person provides administrative and clerical support for the program. Do not, repeat DO NOT, underestimate the importance of this role. This is a critical role for the success of your program. Duties for this person include: setting up meetings, finding meeting rooms, distributing meeting notices and agendas, records meeting minutes, stores meeting documents, maintains member contact list and coordinates RSVPs. This is just a partial list. Hopefully you get the point. This person does just about everything to make sure the

program functions efficiently and effectively. This role requires a professional assistant with experience organizing meetings, especially meetings with Medical Staff.

6) **Data Analyst:** A highly functioning Value Analysis Program must have access to data and someone trained in data analysis. The data must be consistent and whenever possible up-to-date. Bench-marked data is also very useful even when a program is using itself as a bench-mark. Bench-marked data to other healthcare systems can also be very useful. In the beginning, a program must establish a baseline and compare this baseline to itself over time and to others when such comparisons are available. If the goal is performance improvement, the program must know where it stands in order to understand where it needs to go. The Data Analyst will coordinate with the Contracting Lead to develop a list of impactable spend from the contract calendar. Impactable spend is that proportion of supply spend that a Value Analysis Program will target for a given time period.

7) **Research Analyst:** The Research Analyst is that person who will be responsible for researching the products and services being considered by the Value Analysis Program. The Research Analyst will conduct literature reviews and request information from vendors, the FDA and external sources of information (Hayes, ECRI Institute, Healthcare Business Insights, etc.). This person will organize and present their findings at Value Analysis meetings as requested by leadership.

8) **Multidisciplinary Representation:** Credibility and transparency will be extremely important for the success of this program. Identifying formal and informal leaders who work in the hospital and will be in a position to obtain direct feedback on the impact of decisions made by the program will be helpful for the effectiveness of the program. It should be a group who represents physicians and nurses and is very familiar with the operations of your system.

PROCESS

1) **Regular meetings:** If you want your Value Analysis Program to succeed, you will need to have regular Value Analysis Committee meetings. How often is up to the program's leadership. As a general guide, the more contracts you review, especially if they are contracts involving PPIs, the more frequently you will need to meet. Always schedule your meetings far in advance and on a specific day, time and location. (e.g. first Monday of the month at 5:00 PM in Conference Room A) Consistency is important for driving engagement.

2) **Maintain an Agenda and Meeting Minutes:** The agenda is used to drive the discussion and the meeting minutes will document what was discussed. The agenda needs to be shared, in advance, with the Value Analysis Committee membership. Members must play a role in shaping the agenda, and the best way to do this is to provide a summary document of the upcoming annual contracting calendar. The summary document should include the product category, the vendors, the contract expiration dates and a note indicating whether this category is a PPI. Another way to do this it to request agenda items be forwarded to your administrative assistant or co-leads. This communicates to your members that you are seeking their input on how the program will be managed.

3) **Established Reporting Structure:** The Value Analysis Program should report to the hospital's or healthcare systems executive leadership. The model you choose is up to the individual program. There are two basic models of governance. One is a Centralized Model and the other is a Decentralized Model. A Centralized Model is one where the Value Analysis Committee reports up to the executive leadership of the healthcare system. In this model the Value Analysis Program provides this service for the entire system. In a Decentralized Model, the Value Analysis Committee reports directly to the C-Suite. Regardless

of which model you choose the important feature is that the Value Analysis Program is reporting up to leadership. It creates the governance for your committee and provides credibility for it as well. Highly functioning programs will typically report to both the executive leadership and medical leadership. This allows for better communication between Medical Staff and the Administration. We will discuss this topic more in the chapter on accountability.

4) **Communication Plan:** The Value Analysis Program must have a consistent method of communicating important information from the program to its stakeholders, as well as a consistent method of hearing back from them. This can be everything from a monthly newsletter, to a blast email, to regular town hall meetings. It is entirely dependent on the size, complexity and culture of your hospital or healthcare system. Either way it must be regular, consistent, easy to access, easy to interpret and reliable. It typically will come directly from the Value Analysis Committee or its leadership. Most programs will have various formats for communication based on the information being communicated and the intended audience.

5) **Data Review:** One of the first items a Value Analysis Program will need to determine is what metrics it will track to determine if their performance improvement plan is working or not. In the chapter on performance improvement, we will provide a list of metrics. One of the biggest problems a program may encounter is to have data being generated at some level in the organization, and not have a process for accessing the data related to its programs. If your system has an enterprise data warehouse (EDW) this process may already be established. Healthcare systems, like most large companies, will often have multiple data systems with multiple operators managing those systems. An EDW pulls data from all or many of those systems into one data warehouse. This allows for better data management and the ability to analyze data from multiple

sources at once. Systems with an EDW have usually also created processes for pulling this data for analysis. If your system is set up like this, then you will need to tie into this process. If not, then you will inevitably be working with several people who will need to be coordinated for data abstraction and analysis. Focus first on the data you can get. If you are getting push-back from the people who control the data flow, then it is time to work with your executive leadership to find a way to get the data. Once you have determined the data you want and who controls access to it, decide how often your committee needs to see the data. Data must lead to some action so it is important to have enough data to feel comfortable acting on it. To summarize:
   a. Determine the data to be tracked
   b. Develop a process for abstracting and analyzing the data.
   c. Determine the type of data needed to assist with evaluating products and services.
   d. Create a dashboard for data reporting.
6) **Goal Setting:** This will need to be coordinated with #5 above. Have some idea of what goals you will have for your program. Start with an estimate of your impactable spend for the coming year. This is the total value of the contracts that will be expiring in the next year. Depending on the average period of your contracts, this will be about 30-40% of your total annual spend for supplies. Next, target specific contracts you think you will be able to impact. If a supplier has a monopoly on a specific product category, and your utilization is not likely to change, then you may want to focus your efforts on a more impactable product category. On the other hand, if you working with a group of surgeons who are willing to sole source or dual source a product category, and you currently have contracts with eight vendors, then this is a prime target. With the surgeon's support you should be able to obtain a favorable price point in this product category. In addition, goal setting should be realistic.

Making goals such as we will cut supply spend by 50% in the next six months is not realistic. A goal such as, we will reduce the price point for product "x" by 6% in the next contract cycle is more realistic. If you have never used the S.M.A.R.T. approach to goal setting, then now is the time. S.M.A.R.T. stands for specific, measurable, achievable, results-focused and time-bound. (See above example.) Use this or a similar method of setting goals for all of your program's goals.

7) **Work-flow:** In order to be an effective Value Analysis Program you will need to have a defined work-flow. If you are unfamiliar with how this is done, and graphically represented, it will be worth the effort to find someone who can do this for you. Typically, this will be someone with a process engineering background. What they will flow out for you is all of the steps necessary for moving a product or service through your decision making process. For example, the majority of products up for contract renewal will not require the services of the Value Analysis Committee. These products can be reviewed in a different forum and the process expedited. This part of the process will need to be spelled out in specific terms, as well as when a product will need to be reviewed by the Value Analysis Committee. We will discuss this in further detail in the chapter on Accountability.

TECHNOLOGY

1) **Inventory Management System:** Supply Chain Services will have a method for managing its inventory procurement, invoicing and expenses. The software available for doing this has evolved considerably in the past decade. A system that can order and track both contracted and non-contracted items is essential for establishing a supply formulary. A Supply Formulary functions just like a Pharmacy Formulary. Requested items must come from the formulary. Items that are not formulary will require an approval process. When that approval

process is built into the software managing the inventory it is more likely to result in a well-managed formulary. In the past, without a centralized way to manage the inventory it was all too common for local approval processes to undermine the formulary process. Why is this important? Because when everyone gets what they want, whenever they want it, then costs become uncontrollable. Runaway costs are an important reason why hospitals and healthcare systems get into financial trouble.

2) **Electronic Medical Record:** There are two important reasons why an EMR is good to have in a Value Analysis Program. When you have access to an EMR, this creates opportunities to standardize order sets and collect data. Both of these are key to a successful performance improvement program. Standardized order sets allow your program to create care pathways which can use a standardized supplies leading to cost savings. The data generated from your EMR allows your program to monitor whether those care pathways are working, and if they are producing the clinical, utilization and cost improvement goals your program established.

3) **Risk-adjustment and Bench-marking Capability:** There are a number of valid programs available in the market place and many large systems have access to this capability. Administrative data is based on coding, which is dependent on documentation. Healthcare systems with strong documentation programs tend to shine in these types of databases so your external bench marks may be somewhat deceiving compared to these organizations. Do not despair, they can still be useful. First, you can always use them to compare your current data to your historical data. When you implement a new change, you can still track your progress in a before and after manner. If you do not have access to these types of programs you will still need to track your performance. You can do this by tracking dollars saved or number of new contacts impacted with lower price points compared to historical data.

4) **Centralized Storage for Documents:** All documents associated with your Value Analysis Program will need to be stored electronically in a common safe place. Typically, this can be with a SharePoint or some other type of cloud based storage. Members must have access to this repository. This is where you will place documents such as meeting agendas, meeting minutes, dashboards, literature, policy statements regarding supplies, etc. Avoid placing this on someone's (Administrative Assistant) computer. If that person leaves the company, there is the risk that everything you've done for the program could be lost as well.

5) **Tele-conferencing Capability:** There are going to be times when members of your program will need to participate from a remote location. At the very least, programs should have a call-in number for members who can't physically be present. Make sure your rules allow for members to call in remotely. It is also important to have a consistent call-in number and not change it meeting to meeting. You are more likely to get participation when the call-in number and any live-meeting access is consistent. (See chapter on engagement.)

CULTURE

1) **Quality Improvement Program:** All hospitals should have a quality improvement program. The question is, how active is your program. If the only projects your quality department works on are required for legal, compliance or regulatory purposes, then it falls in the category of just above the minimal threshold. Quality departments that are actively working on other projects, in addition to the minimum, are demonstrating a cultural attribute consistent with high performing organizations. If your Quality department is actively working on a Value Analysis quality improvement project, then you are well on the way to success.

2) **Recognition Awards:** Has your hospital or healthcare system received any awards for performance? This may include Truven Top 100 Hospital, Truven Top 15 Healthcare System, Baldridge Performance Excellence Award or any similar type of recognition for performance improvement. If the award is for your Supply Chain Services (Gartner Top 25, ECRI Healthcare Supply Chain Achievement Award, etc.), then this especially important. If so, then there is a good chance you are part of a culture that is going to be very interested in seeing your Value Analysis Program develop. While these awards are not specific to Value Analysis, they are markers of a culture that understands the importance of performance improvement.

3) *Value Analysis Supportive Environment:* Do the members of your program feel supported? Does your administration listen and respond to your program's needs regarding support for your program? Do your departments and Medical Staff listen to and participate in your Value Analysis Teams. If the answer to any of these questions is "no", then your local culture will become a major challenge to the sustainability of your program.

CHAPTER 6
ACCOUNTABILITY

One of the first decisions you will need to make is what type of accountability model your Value Analysis Program will use. This will determine what type of governance your program will have. There are two basic models currently in practice in facilities that have a program. Which one to use is entirely up to the individuals who are charged with developing a program. The two models are the Centralized Model and the Decentralized Model. We will discuss the pros and cons of each model, which you can use in your decision. Following this we will review the process flow between committees. While much of this discussion could easily fit under the heading of infrastructure, this discussion will focus on work-flows and processes that specifically maintain accountability.

The Centralized Model

The Centralized Model for a Value Analysis Program is one in which the program provides oversight for an entire system. This model is most commonly found in healthcare systems that utilize an "operating model" of governance. What this means is that if there are ten hospitals and twenty clinics in the system, then they all report up to a single senior management team, and this team reports to a single Board of Directors. Decision making occurs at the senior management level and decisions are operationalized throughout the system at the unit level (hospital, clinic, skill nursing facility, etc.). There will be one senior leader over Supply Chain Services and this division will be responsible for all decisions regarding supplies for the system. If your system functions in this manner, then this is where you want to plant the flag of your Value Analysis Program. I'm not saying it is a waste of time for individual facilities to have their own Value Analysis Program, I'm pointing out the facts of where the decision making power is in an operating model.

The biggest benefit of this model is scale. Because the system is speaking with one voice in negotiations with vendors, you will have access to much better pricing. Another advantage is vendors prefer to negotiate with the larger systems. It saves them money as well. Instead of sending out teams to meet with multiple hospitals, they send out one team to negotiate for all of them at once. This is the same concept behind GPOs. If a hospital is willing to accept the GPOs pricing, then this saves everyone a lot of time and trouble. The downside of this approach is you are then stuck with whoever contracts with the GPO. You are also stuck with the GPOs pricing. There are going to be times when local contracts (contracts directly between the healthcare system and vendor) are better. This is in part because the GPOs charge a commission. It is also because the GPO is negotiating a price point for all their members. What may be a great price point for a small hospital with little market share, can be a relatively expensive price point for a large integrated delivery network (LIDN) that is a dominant player in its markets.

Another benefit to this model is that if the system has a single inventory management system in place, then it is possible to rotate stock based on system needs. One way to do this is to have a centralized storage facility for stock and distribute it in real time based on facility's needs. Another way is to take back stock at risk of reaching its expiration date from a low volume center and placing it at a high volume center. Another advantage is if the healthcare system controls enough market share, then it may be able to negotiate a consignment deal. This is where the vendor owns the stock and is responsible for maintaining it. Now the hospitals do not have to worry about this at all, and managing the inventory is a vendor responsibility.

So what is the downside of a Centralized Model? If your system does not have an operating model of governance, then it most likely is part of a holding company. A holding company owns the facilities but allows the local management teams and Board of Directors to function relatively independently. Developing a Centralized Model in this governance structure can be challenging. It is not impossible, just challenging compared to an operating model. Most often legal agreements will need to be in place that obligates the facilities to manage their supplies as if part of a single system. As described above there are incentives for doing so, but some may not want to go this route.

If you have a fiercely independent medical staff, who have very strong feelings regarding preference items, you may have some difficulty with following the decisions of a centralized decision making body. This may place the local C-Suite in a difficult position with their medical staff. It is not uncommon for a physician or group of physicians, who contribute a significant amount of a hospital's margin to use this to obtain things they want. It is usually stated explicitly or implicitly as, "If we don't get product 'x', then we are taking all of our cases down the road to your competition." Many a CEO has signed off on such deals just to keep peace in the house. Too much of this, and a Centralized Model will become ineffective. The vendors are highly tuned into this type of behavior and will always exploit it to their advantage. As long as a CEO or CFO are willing to play along with this to protect their facilities margin, this will create a vulnerability for the system. While this may serve short-term quarterly or annual goals, there can be long-term consequences to undisciplined management of supply expenses. There are ways to prevent this to some degree.

It comes back to an inventory management system. If there is only one way for anyone in the system to procure supplies, then an inventory management system can be the gatekeeper. Items in the system are formulary, restricted formulary (specialty item), and non-formulary. Any supply on formulary is the system approved supply item. If a physician orders a non-formulary supply, then it is auto-substituted with the formulary supply, unless the physician or nurse completes a non-formulary request and it is approved. Only formulary and approved non-formulary item invoices will be paid. This prevents the vendor from bringing in supplies for a physician and then invoicing the hospital. This has been a very common practice in the past and was a known backdoor method of selling hospitals non-contracted supplies at list price or more. Inventory management systems are closing this door and forcing vendors to negotiate upfront in good faith.

Decentralized Model

A Decentralized Model for a Value Analysis Program is one where the individual facilities are making decisions for their own facility. They may work jointly with the system or other facilities to gain some scale and leverage with the vendors, but remain independent and for the most part autonomous.

The benefit of this model is therefore related to the value placed on independence and autonomy. While this benefit may seem to pale compared to the advantages of a Centralized Model, it may be the only sustainable model in some hospitals. If your hospital is part of a holding company, you may already have a Board of Directors that prefers to make decisions for itself regarding supplies. Some of this may be cultural, but there can also be some unique characteristics of the community they serve which makes them a more qualified decision making body. Communities can have special needs (elderly population, indigent/homeless population, long distance from urban centers with centralized distribution) and local boards may be in the best position to understand what services and supplies will best meet those needs.

If you have a highly engaged medical staff who prefers to be involved in the types of decisions Value Analysis Programs make, then there will be a trade-off in favor of the engaged medical staff. What benefits you lose in scale can be off-set by a medical staff that is in tight alignment with the administration. When done right, this can be a formidable alliance. It can be so formidable; it makes one wonder why it is not done more often.

So which model is the best one for your program? The best one is the one that works the best for your organization. Both of them have the potential to work. The one that doesn't work very well is always going to be the one that lacks one or more of the seven pillars of successful programs. There are many ways to do it wrong, and only few ways to do it right. Do more of what you know works, and less of what you now know doesn't work.

Work Flow Between Committees

It is impossible for any one team to review every product and service that enters a healthcare system. It takes multiple teams which is why Supply Chain Services is comprised of a small army of trained professionals who do nothing but this all day. A Value Analysis Program is not designed or expected to do the work this group does. It is really focused on a relatively small portion of that work. While this work may seem relatively small in comparison, its impact on patient care and the provider community is much larger. So how do we narrow down what needs to be addressed by a Value Analysis Program's committees? It is starts with having a defined work flow. Refer to the Chart 1 (at the end of this chapter), for the following discussion of this defined work flow.

The Value Analysis Program is really a system of committees that will be reporting up to a Value Analysis Committee. The Value Analysis Committee will provide oversight to the program and should report directly to the senior management of your healthcare system or hospital. There are a number of ways to do this as long as the chain of accountability is maintained within a defined governance structure. Chart 1 is one way to do this. There are also any number of ways to modify it. This will be dependent on the size of the system or hospital where you are implementing a Value Analysis Program, its culture and the current reporting structure.

In this chapter, we will discuss one way to put this together. Your Value Analysis Program will have a Value Analysis Committee that will provide oversight for the program. This committee will be directly reporting to the senior management of your healthcare system or hospital. There are a number of subcommittees who will report to your Value Analysis Committee. The names of each subcommittee are less important. What is important is the function of the committee. The work-flow is dictated by a series of questions. The answers to these questions will determine which committee will be evaluating a particular product. For the purposes of this work-flow, the committee names are: Commodities Committee (CC), Technology Review Committee (TRC), Supply Exception Process Committee (SEPC), Value Analysis Team (VAT), and the Strategy Committee (SC). Each of these committees will be discussed in further detail.

Commodities Committee

Most of the items Supply Chain Services provides are commodities. A commodity is any good or service which cannot be differentiated based on quality, across a given market. Wheat and oil are typical examples. In a hospital setting, this can be most types of dressings, tubing and paper products. The market has long since recognized the commoditization of such products and the differentiator for value s price. So the first question to ask is, "Is this product a commodity item and up for contract renewal?" If it is, then pass this off to a Commodities Committee (CC). This is a group of Supply Chain Services personnel who review and make recommendations regarding this category of supplies. It makes up a large part of the inventory, and Supply Chain Services are already on top of this. They are very good at this, and there is usually no reason to involve providers in most decision making in this category. On rare occasions, there may be times when an item in this category will need to be bumped to one of the other committees, and if so then simply pass it on to the committee established to manage these fallouts.

Technology Review Committee

The next question to answer is, "Is this a request for a new product for an established service line or support function, and is not considered a 'preference item' or have 'strategic' implications?" If the answer to this is yes, then refer it to your Technology Review Committee (TRC). This is a system or facility based committee that reviews all new technology entering the system. This committee may be known in your system by a different name, but there is a good chance that there is someone or some group who already serves this function. There are a number of items that must be checked prior to bringing a supply into a healthcare system's inventory. Things such as, is the item approved by the Food and Drug Administration (FDA)? It doesn't happen often, but it never ceases to amaze me when a vendor, without an FDA approved product tries to lock-up a contract with a hospital prior to FDA approval. Yes, it happens, so make sure you have a process in place to prevent this.

It is good practice to have a standardized process in place for new technology requests. A request form, either in a paper or digital format, with all of the elements (products name, vendor, model number, FDA approval, pricing, etc.) can streamline this process for the committee. Having a requesting physician or other provider complete the form helps in several ways. This person is the one who is best to articulate the specific need for a new product compared to what is already available. Vendors can assist this process, but it is essential to know that there is a clinical need and provider demand for the product. If not, then this committee will be inundated with vendor requests. We made it a requirement that any provider requests also be accompanied with a completed disclosure form of any relevant conflict of interest. With access to the Centers for Medicare and Medicaid Services Open Payments website, checking on these disclosures is much easier. You can also access this information at the Pro Publica website under the "Dollars for Docs" tab. Just enter a physician's name and it will provide the most recent details of payments to physicians from medical device and pharmaceutical companies. Is this really necessary? Yes, because if you review enough physician requests for new products, then you will inevitably encounter a small percentage who will be receiving money from the company whose product is being considered for review. I am amazed at how many physicians "forget" to include this in their disclosure. If a physician is presenting at a medical conference, then it is expected that disclosure is stated upfront. This is the standard, and should be upheld in the product review process as well. It doesn't eliminate the product from review, but is done to inform decision makers of the potential for conflict of interest.

Supply Exception Process Committee

The next question in this work-flow is, "Is this a request for a product which is a physician or clinician preference item?" If so, then the next question to ask is, "Is this a one-time request?" One time requests are important to consider as part of your work-flow and can be referred to the Supply Exception Process Committee (SEPC). It may be for an off-contract, highly specialized orthopedic implant which is custom fitted and literally the only option for a particular patient. These are rare requests, but you will need a process to assess and approve or deny the request. There should also be a request for for this process but it should be simplified. Many of these types of requests are urgent in nature and you will want to review and record your thinking and decision. This is where a physician leader involved in Supply Chain is very valuable. We required that these were reviewed and signed off by both the administrative and medical leader of the Value Analysis Program. Almost all of the time these requests were very reasonable and did not require a formal committee to approve. Having the group who can be available for further discussion is helpful for those cases where it is unclear what the advantages of the requested item is. It is especially important to communicate to the requesting physician that when approved, it is a one-time approval. In order to get the product on the formulary, then it will need to go through the request process to bring the item onto the formulary. When it is not a one-time request, then this will need to be first reviewed by the TRC. If it is determined to be a PPI by the committee then it will need to be reviewed by a Value Analysis Team (VAT).

Value Analysis Team and PPIs

So exactly, what is a PPI? Needless to say, there can be a fine line between what stays in your TRC and what gets forwarded to a VAT. Before we discuss the VAT, lets first define PPI. PPIs are typically specialized supplies that are chosen for use in patient care by physicians based on their familiarity of the product and expectations on the clinical outcome[15]. Physicians tend to be insensitive to the price of the product and in most healthcare environments are not directly responsible for paying for it. PPIs constitute a significant proportion of a healthcare system's supply budget and are therefore frequently targeted by hospitals and vendors for the opportunities this

---

[15] Schneller, E.S., and L. Smeltzer. 2006. *Strategic Management of the Health Care Supply Chain*. San Francisco: Jossey-Bass.

presents. Examples of PPIs include, orthopedic implants, pacemakers, surgical mesh, and endomechanicals to name a few.

Value Analysis Teams can take many forms, but the underlying concept behind any team is that there is some core group who provides the basic functions of meeting logistics, research, data analysis, product review, contracting and strategic development. The remainder of the group is primarily comprised of subject matter experts (SMEs). Large organizations can have multiple teams function in this manner. An alternative approach is to have a single core team and bring in SMEs as needed. The advantage of this approach is the core team develops an expertise in the process of value analysis, and the SMEs are brought into the discussion when their expert advice is needed to guide the decision making process around specific PPIs.

Strategy Committee

The next question in the process is, "Is this a request for a product that impacts strategy or population health?" There are going to be times when there are decisions that will need to be made at the system level. An example of this may be a request for product support from a cardiothoracic surgeon who wants to start-up a heart program at your hospital. The hospital may be part of a system that has a regionalized approach to cardiac surgery, so the request must be made in consideration to the system's strategy for cardiac care and population health for the region. This is where you may need an option for creating another committee. Let's refer to this as the Strategy Committee (SC). Large organizations will most likely have a separate team who will be charged with making these decisions. The Value Analysis Committee (VAC) would be the referring committee in these situations. In smaller healthcare systems and hospitals, it may fall on the shoulders of the VAC. The important thing to remember is, there will always be requests that will impact strategy for the organization. Whether it is a VAC or SC that evaluates this request is less important. The other important thing to remember is that requests of this type coming from a physician should always be considered a PPI. This is one that may have a direct impact on their personal practice and success in your community.

Accountability will be extremely important to the success of your program. Decisions will need to be made, and contracts will need to be signed or not signed. Supplies will need to be available to take care of patients. One of physicians biggest fears is that someone, who is not accountable for the outcomes of patients, will be making decisions about the supplies physicians use to effect patient outcomes. This is usually attributed to "bean counters" who are just trying to find the lowest bid price for supplies. Having worked with many "bean counters" over the years, my personal experience has been just the opposite. Yes, they are always in search of a lower price for a particular product category, but what they are really in search of is the best value. What they often lack is insight regarding the qualitative advantages of one product over the other. It is unfortunate that we've created a system that pits the "bean counters" against the "Docs". One has expertise in product pricing, negotiating, contracting and inventory management. The other has expertise in product utility, quality and clinical outcomes. Value analysis is about bringing these SMEs into the same room in order to identify the best value. They can answer the question, "What is the most cost-effective supply solution for our healthcare system?"

Value Analysis Committee

As previously mentioned the Value Analysis Committee is the executive arm of your Value Analysis Program. Depending on how you structure your program, this committee may be the only one who meets. The committees described in this chapter are therefore going to be specific functions expected of the one committee. While it may defer the evaluation of commodities to Supply Chain Services, it will always need to be ready to receive referrals from this group. In addition, in order for this committee to be effective it must report directly to senior management and preferably to your senior physician leadership. Both administrative and clinical leadership must be aware of your Value Analysis Program and actively support it. How can you tell? If you ask a member of senior management or physician leadership what value analysis is and their response is, "What's that?", then you don't have their support. They don't have to be experts. That's why they hired you. They do need to be able to articulate the value of the program. In the chapter on performance improvement, we will discuss how you will be able to impress upon them this exact point. Before we do that, let's talk about leadership.

Chart 1: Value Analysis Program Work Flow

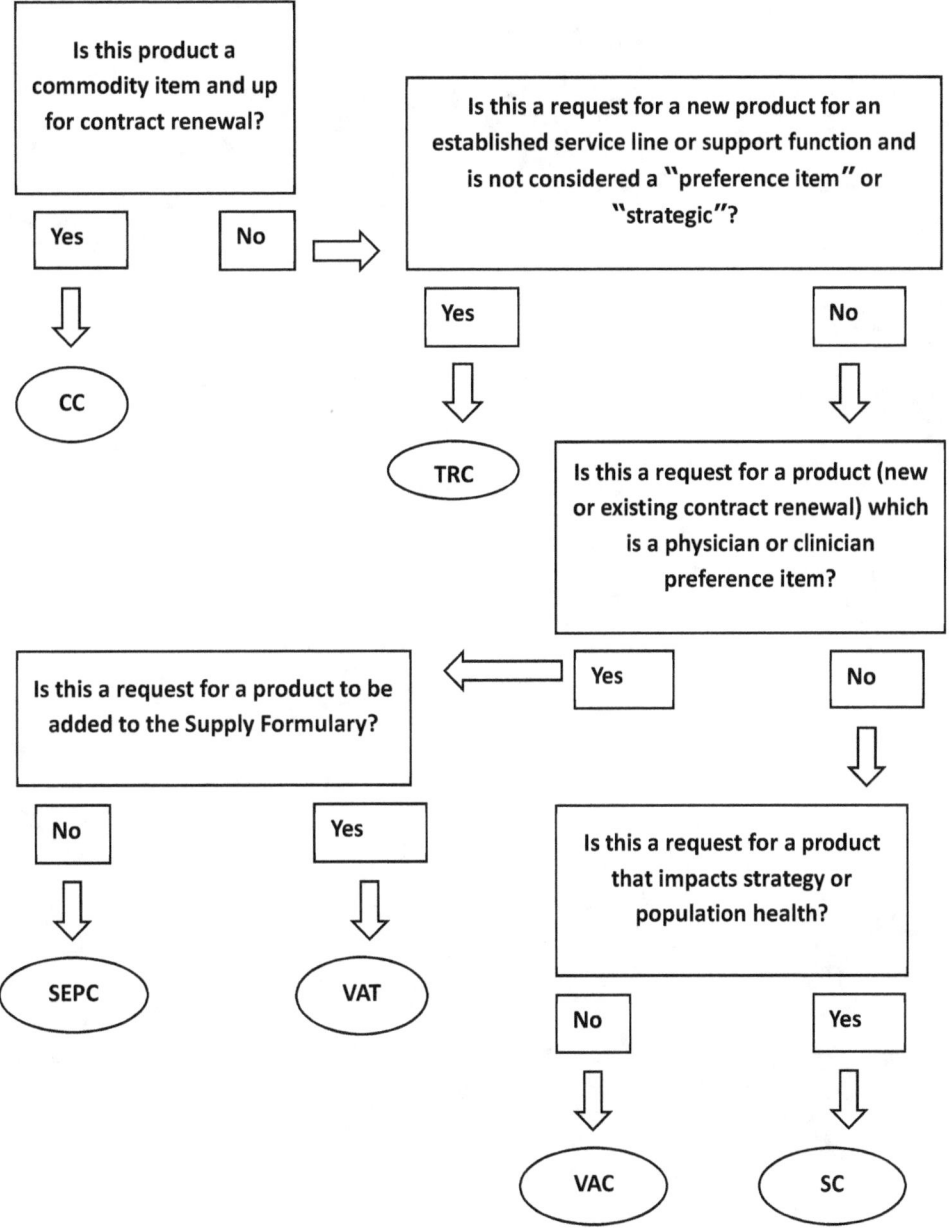

CHAPTER 7
LEADERSHIP

The CEO of a healthcare system were I worked, constantly reminded us that leadership matters. When it comes to developing your Value Analysis Program, or for that matter any program, this is going to be one of the most important things for you to remember. If you have read this far into the book, then guess what? There is a very good chance you are either a leader, or an aspiring leader. Implementing and sustaining your program will be entirely dependent on your leadership. Yes, every other chapter you have read so far is very important, but without leadership, the rest of the seven pillars will not get you very far.

You may find yourself thinking "I'm not ready for this." What you must understand is, this is your fear speaking to you. If others have enough confidence in you to lead your program, then you are probably more than ready. The fear still needs to be confronted. Leadership is personal in the beginning and all leaders must first begin to lead themselves and face their fears. When I was in medical school, I had an immense fear of public speaking. Like many, I was asked to present at a conference once. It was a short patient presentation, followed by a few questions. Routine stuff for a medical student, but not for me. As I stood up to begin my presentation, my knees were shaking so bad, they gave out and I collapsed back into my chair. Fortunately, one of my teammates saw my predicament and quickly jumped up and said, "Terry, you sit down, I've got this covered." I was humiliated. As we were leaving the conference, I swore to myself, that I would never let this happen again. Around this time, I was introduced to a book written by the late Dr. Susan Jeffers, titled *Feel the Fear and Do It Anyway*. The premise of the book is that both successful and unsuccessful people feel fear. The difference is that the successful people confront whatever it is they fear, even in the presence of fear. That book was a great inspiration for me. From then on, I took on any and all opportunities for public speaking in medical school and during my surgical residency. It took me awhile, but I finally conquered the fear of public speaking. There are very few natural leaders. Most of us learn it on the job. If you want to become a leader, then feel the fear and do it anyway. If I can do it, then so can you.

In this chapter we will discuss some of the key takeaways regarding the important behaviors of successful leaders and how to use leadership to maximize the effectiveness of your Value Analysis Program.

How can something so important as leadership seem so elusive? Part of the reason is because leadership means different things to different people. While there are certain qualities that most people will want to see in their leader (courage, integrity, calm under pressure, etc.), there are expectations of a leader that can vary significantly depending on the stakeholder group with whom they may be interacting. Awareness of those expectations and understanding why they are important is half the battle. We'll look at each of the major stakeholder groups with a particular focus on what they will expect of you as a leader and why.

Physicians

A few years ago I was addressing our Chief Medical Officer (CMO) meeting about a particular supply chain issue. I was looking for their support, and assumed I had it before I walked into the room. It became obvious, in just a few short minutes after I began the presentation, that I was not going to get their support. Afterword a colleague approached me and gave me some advice. He said it sounded like I was speaking to them as CMOs, and as part of their facility's administration, which was true. What I forgot was that all of them are physicians first, and I did not speak to them from this perspective. He was right. Physicians, regardless of title, will always need to be addressed as physicians first. So what do I mean by this?

Physicians value what is right for patient care and the scientific method. This is how they were trained, and it is what they know and understand. Approaching them with the business case for using a particular supply over another is probably not going to win you many friends with a physician audience. You may be able to make this case to the hospital administration, but, as a general rule, it will fall on deaf ears with a physician. They need to know from the beginning that quality and patient safety is the overriding concern. They understand the importance of improving utilization and reducing cost for the hospital, they just need assurance that utilization and cost improvements are not being achieved by subordinating quality and safety for patients and their community.

There is also something else that is very near and dear to physicians. As a group physicians value autonomy and independence. In particular, they want to be able to practice without the sense that everything they do needs to be micromanaged by some outside entity. It is not that they do not understand the importance of guidelines, best practice and the merits of standardization. They do. They want a say in how they will be expected to practice and clear expectations on performance. Just as words like "compliance", "policy" and "mandates" tend to be disengaging to a physician, leaders who do not listen to what physicians are saying, and do not articulate performance expectations will, themselves, become disengaging to physicians.

Leaders who are attempting to appease one stakeholder group (hospital administration) while addressing another stakeholder group (physicians) will quickly find themselves fighting a losing battle. The last thing you, as a leader, need is to be perceived as a puppet of the administration. It is okay to focus your team on quality outcomes. If you can do this, then the road to improving utilization and cost will be much easier. This is for two reasons. The first reason is because if your physician group knows you are doing what you are doing for patient care, then they are more willing to work with you on other issues such as utilization and cost. The other reason is because excellent quality inevitably leads to better utilization and total cost of care savings.

In the supply chain world, there will be one area that will prove challenging, and that is "physician preference items" or PPIs for short. Unfortunately, there is not a great deal of good evidence comparing one type of medical device to another in terms of patient outcomes. Historically physicians have used their personal experience as a guide to differentiate products. This is why they are referred to as "preference" items. When there is data suggesting two products are comparable, it is not uncommon for a physician (okay usually a surgeon) to counter with, "But in my hands, the product I use is better than the other." Usually this is the product that costs more. Did you ever wonder why this is?

Physicians, particularly surgeons, are trained with a variety of products, which I will refer to as tools. Over many years, a surgeon develops a strong sense of comfort with particular tools. The tool feels a certain way in their hands. When surgeons operate, they operate in a box. Oftentimes the operating room is referred to as "the box", but there is another box where a surgeon operates. It is a relatively small area within our focus of vision. It may be no more than about eight by eight inches in size. Sometimes it is smaller and sometimes it is larger but the point is a surgeon's focus is in the box. That box is where the operation is occurring within the patient's body. We are trained not to take our eyes off this area. When a surgeon asks for a tool, it is handed to them by the surgical scrub technician, all while they are looking in the box. The instant the tool is pressed into a surgeon's hand, he or she will know if it is the right tool without looking at it. If it is not the one a surgeon normally uses, then it creates a sense of anxiety. The surgeon is now distracted and needs to look outside the "box". If the surgeon has never used that particular product, they will now have to familiarize themselves with it. This all takes precious time, and throughout that time the anxiety continues to build and erode at their confidence. This is not a state of mind we want surgeons operating. It can be avoided.

I bring this up to help you, as a leader, understand something that is fundamental to anyone who uses tools. There is a connection, at an emotional level, that surgeons, and all proceduralists have with their tools. To refer to it as, simply a preference, trivializes the connection. The medical device manufacturers understand this connection, and work very hard to maintain it with their products. As the physician lead for value analysis, you will need to take this into consideration.

Physicians are willing to try new products. In fact, they do it all the time. What the vendors figured out, a long time ago, is there is a right way to do this and a wrong way. Signing a contract for a lower priced product and abruptly replacing the old product for the new is the wrong way. Having a surgeon involved in the decision to try a new product, and then getting the new product into a surgeon's hands in a controlled, elective environment is a better way to do this. They need to familiarize themselves with how the product works. This can begin in the operating room lounge and then slowly introduced into a surgeon's easier cases. There may need to be a transition period anytime you are moving from one vendor to another in the same product category. This could be very quickly for simpler items (suture) and weeks to months for more complex items (orthopedic implants). If this process moves too quickly, you may experience a backlash. You as a leader will be in the firing line when this happens. Understanding the surgeon's point of view will go a long way in managing the issues that develop when changing out "preference items".

Administration

Not only will different stakeholder groups have varying expectations of you as a leader, but don't be surprised if individuals within those stakeholder groups also have varying expectations of you. Early in a new role as a physician executive, I met with multiple leaders across the various regions of our organization. I had two questions for each of them. What are your immediate expectations of me, and how would you define success for me in this role, one year from now? All of them gave the same vague responses such as, "more physician engagement", show "courage", "influence physicians" and my favorite, "get doctors to do more of what we want them to do". Needless to say these are not only vague responses, but also unrealistic because they are not well defined. I pressed them further for specific metrics or at least some type of semi-quantified definitions. This is when it got interesting. Each person had a different answer. They were using the same words but meant something different when they used it. I wish I could say this problem was confined to administrative leaders. It is not. I found the same problem with physician leaders. There is a solution for this problem that is not dependent on the other person.

Define success for them. When an administrator says they want to see more engagement from physicians as a result of your leadership on the Value Analysis Committee, then tell them what this means to you, and get them to agree to that definition. For example, you can say that your first year goals for the Value Analysis Committee are to define a quality, utilization and cost savings goal for the team and realize a five percent improvement in each. Then ask the administrator if that would qualify as more engagement from physicians? It is difficult for them to not agree with this because there is a very good chance that no one else has done this or suggested this. It is important for you to think about these promises before you enter into these types of discussions. You want to set achievable goals for yourself. So what makes this an attractive offer to an administrator? It appeals to something they value.

The administration is charged with running the operations of the hospital or healthcare system. They cannot ignore business operations. It is their job, and it is what they are trained to do. Their leadership will be holding them accountable to this. Any discussion you have with them must take this into consideration. In the example above, you will notice that goals would be set for quality, utilization and cost savings. It doesn't say what order these goals will be set. When you meet with your committee you can remind them that you will be setting these three types of goals, but you can focus first on quality. If your surgeons are concerned with the large amount of open and unused supply items they see during a typical case, then you can use that as your first utilization goal. The idea is to introduce goals as a way for the various stakeholder groups to achieve outcomes that are important to them and the organization.

Your administration may have specific goals for you to achieve. In this case it will be important to prioritize these goals and set realistic expectations. Use this as an opportunity to shape what their expectation of success for your role will look like. Programs take time to build, and great outcomes will take even longer. If you are being asked to reduce supply cost by thirty percent, then this is a time to speak up and set realistic expectations. Most of you will encounter reasonable administrators who will work with you. They expect you to take the lead on this, so be brave and just do it. It helps to network[16] with other Value Analysis Committee leads and have some understanding of the types of goals you can accomplish in the first year. It may be as simple as establishing a dashboard to track your performance.

Nursing

Regardless of whether you choose a Centralized Model or a Decentralized Model, you will most likely include a Value Analysis Coordinator at your meetings. This individual will either be your Nursing Lead or work closely with your Nursing Lead and be the primary liaison with the nursing staff. The expectations of the Nursing Staff will lie somewhere between that of the physicians and the administration. While they are trained from a clinical perspective to put the patient's interest first, they will also be employees of the hospital. If they are in management, then there is a good chance that person will be responsible for a budget and may have utilization and cost savings goals for their department. There is nothing wrong with this, but you will need to be mindful of this when choosing your annual goals. The closer you are aligned with their goals, and what they are being incentivized to achieve, the greater the likelihood for your success.

There is another aspect of leadership that will be expected of you. This has to do with conflict resolution. It is not that the other stakeholder groups will not expect this of you, it is just that the group that will expect it the most will be Nursing. This can happen in many ways but the most frequent way is when there is a conflict with a physician. As a surgeon it breaks my heart to say this, but it will most commonly be with a surgeon. It's easy to pass this off to the Department Chair, Chief of Staff or Chief Medical Officer, but all eyes will be on you. If you are a physician or nursing lead over Value

---

[16] There is a group called Physicians for Supply Chain Excellence you can join through LinkedIn.

Analysis and there is a conflict in this area, you will be expected, at the very least, to be aware of it and in many cases to directly intervene. There may be a tendency to shy away from this role but you really should not for three reasons. First, all eyes are watching you and want to see how you respond as a leader. Second, if you can handle this well it is a great opportunity to gain the respect of your team. Finally, as a Leader of Value Analysis, you are the one person who is most likely to understand the issues that are responsible for the conflict. If it is happening with one person, then if may be an issue that is affecting the entire program. In that case, you really are the one person who should be involved.

Value Analysis Committee

This is the group that will be your biggest challenge. The reason is because it is usually composed of members of all the other stakeholder groups. There will always be competing priorities and agendas in the room even in the most organized programs. Identify a common agenda for the team during your first meetings. Data is usually an item of interest for everyone. You will want to know what data you can track, what data you need to track and how your team will obtain access to a reliable data source. Risk adjusted and bench-marked data is always nice to have, but not essential in the early days of a program. As your program grows in strength, communication with the committee members will become important. They are your ambassadors. With growth will come issues of quality and cost, so make sure your program plans on having access to this type of data when the time comes. As you can imagine, if you are going to be having discussions around quality, then risk adjusted data will become very important to your physicians. Be their advocate. They will not let this issue die, and neither should you.

One last bit of advice I've adapted from a column titled "Leader Time: 8 things every CEO must know (from memory)" written by Peter DeMarco. [17]

1) **People:** Know everyone by name on your Value Analysis Committee, and know or be familiar with every person on the subcommittees.
2) **Prices:** What are you charging for your products and services? Is it competitive with the system down the road? You may not think this is important, but once you become a leader people will expect you to, at least, have some idea of this.
3) **Pay:** It is probably not important to know the exact compensation package everyone on your team has, but it is important for you to have some idea if your rates are competitive in the market. The last thing you want to do is to train a top notch value analysis team and have them jump to the competition because the hospital did not have a competitive compensation and benefits package.
4) **Purchasing:** This includes the equipment, instruments, accessories and supplies. You don't need know how much each Band-Aid costs but you should be familiar with all of the higher priced items. Also, have some idea of what alternative supplies cost such as reprocessed equipment when applicable.
5) **Presentation:** Be familiar with how your program is being presented to the physician and nursing community.
6) **Perception:** How is your program being perceived by your stakeholders? A great way to do this is to simply ask your colleagues. Another way is to send out a survey. Don't overdo it. A few questions once a year is all it takes.
7) **Performance:** You must be familiar with your metrics and should know what your current dashboard shows. You will also want to know about significant changes in your metrics. When

---

[17] DeMarco, Peter. Leader Time: 8 things every CEO must know (from memory). The Business Journals. March 1, 2016.

you see improvement, then feel confident and brag about it. Your team expects you to do this. Trumpet the team's success. It is a big booster for morale.

8) **Priorities:** The program's priorities will change as the program matures. This is normal. You will have to constantly be reassessing your priorities with guidance from your stakeholder groups.

CHAPTER 8
PERFORMANCE IMPROVEMENT

In healthcare systems performance improvement (PI) is usually associated with the quality department. In actuality, PI should be embedded in every department and program in the system. If the goal of a Value Analysis Program is to identify the most cost-effective supply solution, then PI will be an essential tool for achieving this goal. In this chapter, we will discuss two types of PI, metrics you can use to populate your dashboard and various strategies your value analysis team can use.

Program Performance

Your Value Analysis Program will constantly be required to prove its performance. Building this into your program from the first day will be much easier than trying to prove it in a post-hoc manner. It also keeps your team focused and engaged when they can see the program's accomplishments in real time. In order to achieve accomplishments, the program must have goals. When you first start working in this environment, prioritizing goals can be challenging. It always seems like there is more work than you can possibly do. That is because there is more work than you can possibly do. Welcome to leadership. You may feel overwhelmed at first. Take comfort in the fact that others have been down this path before, and there are ways to focus your program and streamline the workload. This takes some preparation but once you have it up and running it will be easier with time.

Start with your contract calendar. Contracts are time bound and therefore can be used to queue up the work. Most hospital supplies will have contracts that are one to three years in length. On average they will be about two years in length. The hospitals and the vendors are incentivized to maintain these contract calendars and will only deviate from them under extraordinary conditions. What this means is that for any given year there will only be a certain number of contracts up for review. Depending on how your Supply Chain Services has structured its contract calendar you may be reviewing 30-50% of the contracts in a given year. If the time horizon for your program is one year, then it will be those contracts where you will want to start your program. If you add up the cost of all contracts, then this will be your total supply spend. The cost of all the contracts up for review is therefore the impactable supply spend, and the exact percentage it will comprise of the total supply spend is dependent on how your contract calendar is structured. It is important to understand this because it will determine the types of goals you will set for the program.

Now that you have your impactable spend list of contracts, it is time to start prioritizing your work. A simple way to do this is to break the list up into groups based on the subcommittees discussed in chapter 6 on Accountability. By doing this your list will be narrowed considerably. Most of the list will be commodities. A smaller proportion will need to be reviewed by the Technology Review Committee. The remaining contracts will need to be reviewed to see if committing to a full value analysis is warranted.

There are many ways to further break this down, but the easiest will be to sort the list from largest spend to smallest spend. The top of the list should not surprise you. It will most likely be comprised of implantable medical devices (e.g. hips, knees, pacemakers, etc.). While it may seem tempting to circle the top ten as your goals for the year, it will be helpful to discuss each with your Supply Chain Services team. They will provide a wealth of background information regarding those contracts. They may have received a great price reduction on the previous contract from one of the vendors, and it is unlikely you will get that much in the coming year. It could also be that one of the vendors lower in the list has several products going off patent and will be experiencing formidable competition in the next year. There is a story behind every contract, and your Supply Chain team should know all the details. The goal of this review is to create a list of potential contracts that you will be reviewing with a team of subject matter experts. Once you have this list, set is aside for now. There is another list you will need to create.

Contracting is one way to identify the most cost-effective supply solution. There is another way, and this one really will require stakeholder support. This is through utilization. Let's use products "x" and "y" to illustrate the difference between value created by contracting and that created by utilization. If we have an equal (50%) mix of both products and get a 10% reduction in price, then the 10% savings is pure contract savings. If we have an equal (50%) mix and the price of product "x" is half of product "y", then changing the mix and moving usage to product "x" creates pure utilization savings. Another way to obtain utilization savings is to reduce the overall use of either product on a per case basis. It could be that two product "x's" were once used on every case but new research shows you only need to use one.

Since utilization is another way to create value, you will need to identify opportunities. While this may seem daunting in scale, we discovered a useful way to streamline this process. We created an "Idea Center". We asked all of our healthcare system's teams to look for utilization ideas in all departments. Any utilization idea could be submitted. Supply Chain Services screened all of the ideas, and working with the teams, created a process for implementing those ideas across the system. An idea that could save a few hundred dollars in one small department could end up saving tens and even hundreds of thousands of dollars across a large system. In the first year we generated 472 ideas. By the second year, we generated 938 ideas. The percentage of utilization savings increased from 33% in the first year to 44% in the second year. This translated to about 15 million dollars in supply savings by the end of year two. What was interesting was we did not offer specific rewards for coming up with ideas. People did this on their own for two main reasons. They received recognition for their efforts from their peers, and felt a sense of personal accomplishment for contributing to improving their workplace environment.

There are two approaches to setting overall goals. One is a top down approach and the other is a bottom up approach. In the top down approach the leadership of the organization looks at the total budget and assigns a percentage or dollar amount of it to Supply Chain Services. How this gets apportioned makes for some interesting discussion but suffice it to say, there will usually be a dollar amount assigned to supply savings. Supply Chain then looks at its impactable spend and determines where the opportunities exist that will get them to the budgeted supply savings amount. The bottom up approach begins with the impactable spend list and estimates the potential savings of each contract. The sum of this savings or some percentage of it becomes the program goal. In the early days, there is a great deal of uncertainty regarding the potential impact of a Value Analysis Program so you are better off with a bottom up approach. In time, senior management will come to appreciate the contributions of the program, so don't be surprised when goals start being generated in a top down approach. It is the reward you get for becoming a successful program. Congratulations, you are no longer an experiment; you are officially part of the budget.

Project Performance

While it is essential to have program performance goals it is also necessary to have project performance goals. You can approach this many different ways. For the purposes of this book, we will use the SMART approach. SMART is an acronym which can be used as a guide for achieving your goals. SMART stands for Specific, Measurable, Achievable, Realistic and Time bound. This can be applied to either individual contract projects or utilization projects. As an example we will this with a contract project.

This is most likely to impact a product category which is a physician preference item. Let's say you've identified an opportunity for a specific type of orthopedic implant. There are five vendors providing this product and all five contracts are set to expire in the coming year. Your research shows that your system is paying a higher than expected price for the implant. Estimates are that it is about 10% higher than fair market value for an organization of your size and market share. Two of the vendors have indicated they are willing to reduce the price by as much as 10% if you are willing to agree to a dual source contract at 90% market share. Currently there are twenty orthopedic surgeons in your system and 75% already use implants provided by these two vendors. There are a lot of moving parts in the above scenario. What does this all mean? Is this a good deal? Like all complicated scenarios, it depends.

Let's start with a SMART goal approach. What is your specific goal? It is probably something like, we would like to reduce the price of these orthopedic implants to a fair market value. Is it measureable? Yes, it can be stated as, we would like to reduce the average price of these orthopedic implants by 10%. Is it achievable? Yes, since the amount is within fair market value and you have at least two vendors who are willing to discuss this with you even before the official negotiations have begun. Is it realistic? This is unknown. It is time bound? Yes, since the plan would be to achieve this with the next contract cycle and with this product category.

Not bad, but you can see how the whole project can be thrown off by one simple question. Is it realistic? At the very least you will need to convert five orthopedic surgeons to another vendor before agreeing to the previous offer. What if the 75% of surgeons who use those devices only make up 50% of the total device usage in this category. That means those five surgeons account for 50% of the usage and revenue in this category. How will your CEO feel when he finds out you just alienated the five busiest orthopedic surgeons in the system, and now they are moving their practice to the competition. All is not lost. This is still a great opportunity. We haven't brought any of the surgeons in on the discussion, which is why you will need to plan for a value analysis team to evaluate this product category if you still want to tackle this project. If this is your first attempt at value analysis, then I wouldn't blame you if you wanted to defer this one for the next contract cycle. Sometimes it is just smarter to "run away to fight again some other day." If you were going to take on the project, then how could you approach it.

With regards to a project goal you are almost all the way there. There is some information you will need to be relatively certain that this is realistic. First off, start planning as early as six months in advance. Start collecting data on usage if you don't already have it. If you have it, then continue collecting it because when your Value Analysis Team (VAT) meets you will want the most recent data. Research the implants. Look for evidence based articles. If you have access to resources such as Hayes or the ECRI Institute reports, then pull those. You can also put requests, for this type of information, from these sources, and they will pull the evidence for you. Look at your usage data, and if it is true that five surgeons do 50% of the cases with this product, then start talking to them. Let them know your intention to bring together a VAT, and that you will be inviting them and other orthopedic surgeons who use these implants. The core team for the VAT should start creating a project calendar. This is a calendar that outlines the specific milestones for achieving this project's goals. The earlier you can involve the surgeons, the better it will be. If you are worried they will talk to the vendors, then stop worrying. They will be talking to the vendors. That's okay. They already talk, just about every day.

Schedule two meetings with the surgeons to discuss the contracting options available. Begin the discussion with a conversation on what we can do, from a supply perspective, to maintain or improve the quality of their work. It is always helpful to circulate the results of any research performed prior to the meeting. If the research does not demonstrate a clear qualitative reason for one of the implants over the other, then ask them to comment on why they personally use the product they use. Be prepared for everyone to boast about their great outcomes. If you have outcome data prior the meeting, it is good to share it with them in private. This usually keeps most people grounded. What you are most likely going to hear is a conversation on why each "prefers" a particular product. Remember, forget what the two vendors offered. Your goal is a fair market price with the next contract. It is important to make sure during the course of this meeting to point out that your currently not getting a fair market price for these implants. You also need to explain to them that you think it is reasonable for your system to obtain these products at the same price everyone else is getting. Appeal to their sense of fairness (Rule #3) and request ideas from them on how to achieve this goal (Rule #6). In order to do this, they will need to know the options. This why you may need to schedule two meetings. It may take further discussion and they may have questions or need information that is not readily available at the first meeting. The first time we did this we ended up with an all-play strategy using the capped pricing model. You never know where these discussions will go, but if you provide good information to a highly educated and engaged group of professionals they will always manage to surprise you.

Metrics

The following is a list of metrics to consider for your program. It is not necessary to track all of them, but to identify metrics which are important to your program and supports your organizations overall strategy and goals. While the emphasis is on supply savings, it is always a good idea to be tracking quality outcomes, when the data is available, for specific projects where quality outcomes are a concern.

1) Time to process product request (days)
2) Product requests (#)
3) Product requests accepted on formulary (# and %)
4) Product requests accepted on restricted formulary (# and %)
5) Product requests denied (# and %)
6) Product requests, one time exceptions (#)
7) Product requests, one time exceptions approved (# and %)
8) Product requests, one time exceptions denied (# and %)
9) Ideas submitted to Idea Center (#)
10) Ideas submitted to Idea Center approved (# and %)
11) Ideas submitted to Idea Center denied (# and %)
12) Total supply savings ($)
13) Total supply savings from contracts ($ and %)
14) Total supply savings from utilization ($ and %)
15) Average supply savings/contract ($ and %)
16) Average supply cost/weighted adjusted admission
17) Average supply cost/weighted adjusted day
18) Projects/year (#)
19) Projects achieving goal/year (# and %)
20) Projects completed prior to contract expiration (# and %)

21) Projects not completed prior to contract expiration (# and %)
22) Cost of Value Analysis Program/year ($)
23) Total supply savings from Value Analysis Program ($)
24) Return on investment/year (Total Supply Savings/Cost of Program)

Final Thoughts on Performance Improvement

Each project will contribute to the programs performance. You will have some hits and some misses along the way. Performance improvement is about learning from the successes and failures. Even your successes will be less than perfect. Always ask, "What could we have done differently?" It will take time to develop this program. How long? Based on just the contract calendar it will probably be a minimum of 3 years before you performed a simple review of all the significant contracts in your system. You will not be able to throw everything you have at all the contracts during this time period, and implement all of the utilization ideas you will receive. There is too much work and all of it will not be accomplished in that short period of time. Your program will improve with time, and you will get better at it with each passing day. You will always be working on performance improvement as long as there are new products, technologies and services becoming available in healthcare. Think of it as job security.

APPENDIX
COMMITTEE CHECKLIST

By now you have probably figured out, this is not just a book on how to start-up a successful Value Analysis Program. It is also an introduction on how to start-up successful programs in general. Each program will have specific features which will distinguish it from other programs. The common features are core elements that every program will need to be successful. I've organized these core features into "Seven Pillars". After years of starting different types of programs, the presence of these seven pillars became an unmistakable pattern in the most successful programs. As you develop your Value Analysis Program look for their presence. If you are struggling, then re-examine the program, and see if you are adhering to the elements outlined in each chapter.

All successful outcomes are the result of a series of process steps. Some must occur before others, and there will be some that will not work in all environments. In this chapter a recommended checklist will be provided. The goal is to take you through the planning, the launch of your first Value Analysis Committee meeting and establishing a performance improvement program. This whole process can be done in less than ninety days. A more aggressive timeline can be a done based on your system's priorities. This checklist will take you through the first three meetings. It should work if you plan on meeting every month, every other month or every quarter.

After establishing the Leads and Administrative Assistant, each step will be followed by a recommended responsible person. This will be in parenthesis and will be bolded. Attempt to achieve each item in the checklist, as best you can, in order to launch your program. During this process, keep track of what was essential and what was not. Check the box next to essential items and cross off the list those items found to be non-essential. For example, it may not be customary to provide refreshments at meetings in your organization, then cross this off the list. While the list may seem imposing at first, it is designed to help you develop good habits. It is these good habits that you will want to hardwire into your program. By doing so, you will hardwire barriers to disengaging behaviors. The more effectively you can do this, the more time you will have to focus on engaging your team and building a successful program.

The Value Analysis Committee will be the lead executive forum for value analysis in your organization. Its role is essential for a high performing program. Setting it up correctly from the beginning will allow you reach a higher level of performance much earlier in the process. It requires a lot of "front end" work as they say. Regardless of whether your committee will be meeting every month, every other month or every quarter, plan on meeting at least once a month until you have developed several Value Analysis Team projects. It is also important to identify your program's metrics and begin to track its performance. This will be the basis for your performance improvement program. The checklist is set up to get you there in three months. You may elect to do it quicker but really avoid taking any longer than three months. Create a sense of urgency and capitalize on the momentum to start-up your program. It will be worth it.

I wish you the best of luck in your program's success. If you have any comments or questions, then feel free to contact me at:

TerryLoftusMD@gmail.com or through

www.LoftusHealth.com

Planning for the Value Analysis Committee Meeting

- ☐ Identify a Physician Lead.
- ☐ Identify a Supply Chain Services Lead.
- ☐ Identify an Administrative Assistant. (AA)
- ☐ Set up first planning meeting. **(AA)**
- ☐ Physician & Supply Chain Leads to meet with stakeholder leadership to determine acceptance of Value Analysis Program and which model (Centralized vs. Decentralized) to use. **(Leads)**
- ☐ Stakeholder leadership list.
    - o Chief of Staff
    - o Department Chairs (Surgery, OB/Gyn, Urology, Cardiovascular, ENT, Anesthesia)
    - o CEO of hospital
    - o CMO of hospital
    - o CNO of hospital
    - o Perioperative Services Director
    - o Supply Chain Services Director
- ☐ Confirm type of model to use (Centralized vs. Decentralized). **(Leads)**
- ☐ Identify membership for Value Analysis Committee. **(Leads)**
- ☐ Value Analysis Committee core membership list and contact information.
    - o Physician Lead
    - o Supply Chain Services Lead
    - o Physician Representatives
    - o Nursing Representatives
    - o Administrative Representative
    - o Administrative Assistant
    - o Value Analysis Coordinator
    - o Research Representative
    - o Data Analyst
    - o Contract Specialist (Supply Chain Services)

☐ Set up Prep meeting for first committee meeting. **(AA)**

Prep Meeting

☐ Establish calendar of meeting dates. **(Leads & AA)**
☐ Confirm meetings are consistent (e.g. first Thursday of the Month at 5:00 PM in Boardroom. **(Leads & AA)**
☐ Determine if teleconference capability needed. **(Leads & AA)**
☐ Develop list of membership with contact information. **(Leads & AA)**
☐ Develop meeting agenda template. **(Leads & AA)**
☐ Determine if any documents will be needed for first meeting. **(Leads & AA)**
☐ Determine purpose of committee. **(Leads)**
☐ Develop first meeting's agenda to include: **(Leads & AA)**
- Name of committee
- Purpose of meeting
- Type of meeting
- Date of meeting
- Time of meeting
- Location of meeting
- Call in number with any code (if any)
- Web link for remote conferencing (if any)
- Members names and role for attendance
- Note taker for meeting minutes
- Time keeper
- Note for any member preparation for meeting
- Call meeting to order
- Call to review previous meeting minutes
- Action items
- Responsible party for action items
- Due date for action items
- Agenda items
- Presenter for agenda items

- Time allotted for each item
- New action items
- Responsible party for new action items
- Due date for new action items
- Call for new agenda items for next meeting
- Reminder of next meeting date, time location
- ☐ Contract Specialist to develop contract calendar
- ☐ Confirm centralized repository for program documents.
- ☐ Confirm membership has access to centralized repository.

Prior to First Value Analysis Committee Meeting

- ☐ Confirm next meeting and teleconference capability. **(AA)**
- ☐ Confirm availability of audio-visual equipment and support contact information in case of problems. **(AA)**
- ☐ Send out next meeting notice to membership along with any specific instructions on how to access the meeting via phone or internet. **(AA)**
- ☐ Leadership to review contract calendar for opportunities. **(Leads, Contract Specialist)**
- ☐ Leadership to confirm final agenda. **(Leads)**
- ☐ Send out follow-up meeting notice along with final agenda and any documents to membership. **(AA)**
- ☐ Confirm RSVPs for meeting. **(AA)**
- ☐ Leadership to contact any membership non-responders to meeting notice to confirm RSVP. **(Leads)**
- ☐ Leadership to contact AA with any confirmed RSVP from non-responders. **(Leads)**
- ☐ Arrange refreshments (if any) for next meeting based on RSVP response. **(AA)**
- ☐ Prepare any paper documents needed for membership at the next meeting (agenda and documents). **(AA)**

First Value Analysis Committee Meeting

- ☐ Arrive 10 – 15 minutes before meeting begins. **(Leads & AA)**

- ☐ Set-up any audio-visual equipment and test. **(Leads or AA)**
- ☐ Set-up any teleconferencing equipment and test. **(Leads or AA)**
- ☐ Call in to any bridge line or web enabled teleconferencing 3 – 5 minutes before meeting begins. **(Leads or AA)**
- ☐ Remind call in members that meeting will begin promptly at designated time after making connection. **(Leads or AA)**
- ☐ Call meeting to order at designated time. **(Leads)**
- ☐ If meeting needs to begin later, notify members why meeting is delayed and expected start time. Repeat this every 2-3 minutes until meeting begins. **(Leads)**
- ☐ Take attendance. **(Leads or AA)**
- ☐ Review purpose of the meeting with membership. **(Leads)**
- ☐ Introduce new members and guests. **(Leads)**
- ☐ Remind members to disclose any pertinent potential conflict of interest during any meeting deliberations. **(Leads)**
- ☐ Work through agenda with intent to stick to allotted time for each agenda item. **(Leads)**
- ☐ Review contract calendar for next year and discuss opportunities for Value Analysis. **(Leads, Contract Specialist)**
- ☐ If extra time needed for additional discussion of an agenda item, then ask members if it is okay to go past the allotted time. If so, then notify members of what the additional time limit will be. **(Leads)**
- ☐ Consider creating a Metrics Team to develop a list of potential metrics, and a mock-up of a dashboard for the next meeting. **(Leads)**
- ☐ Request any future agenda items to be forwarded to the Leads and the AA. **(Leads)**
- ☐ Announce next meeting date, time and location. **(Leads or AA)**
- ☐ Adjourn meeting on time. **(Leads)**

Prior to Second Value Analysis Committee Meeting

- ☐ Confirm next meeting and teleconference capability. **(AA)**
- ☐ Confirm availability of audio-visual equipment and support contact information in case of problems. **(AA)**
- ☐ Send out next meeting notice to membership along with any specific instructions on how to access the meeting via phone or internet. **(AA)**
- ☐ Schedule meeting with Metrics Team to determine potential metrics and mock-up of dashboard. **(AA)**
- ☐ Meet with Metrics team to finalize list of potential metrics. **(Leads)**
- ☐ Narrow list of contract calendar opportunities **(Leads)**
- ☐ Pull any data, graphs or charts needed for meeting and forward to Leads. **(Data Analyst)**
- ☐ Leadership to review any data, graphs, charts or documents for next meeting and forward to AA. **(Leads)**
- ☐ Leadership to develop, confirm and forward final agenda to AA. **(Leads)**
- ☐ Send out follow-up meeting notice along with final agenda and any documents to membership. **(AA)**
- ☐ Confirm RSVPs for meeting. **(AA)**
- ☐ AA to notify Leads of any non-responders to RSVP. **(AA)**
- ☐ Leadership to contact any membership non-responders to meeting notice to confirm RSVP. **(Leads)**
- ☐ Leadership to contact AA with any confirmed RSVP from non-responders. **(Leads)**
- ☐ Arrange refreshments (if any) for next meeting based on RSVP response. **(AA)**
- ☐ Prepare any paper documents needed for membership at the next meeting (agenda and documents). **(AA)**

Second Value Analysis Committee Meeting

- ☐ Arrive 10 – 15 minutes before meeting begins. **(Leads & AA)**

- ☐ Set-up any audio-visual equipment and test. **(Leads or AA)**
- ☐ Set-up any teleconferencing equipment and test. **(Leads or AA)**
- ☐ Call in to any bridge line or web enabled teleconferencing 3 – 5 minutes before meeting begins. **(Leads or AA)**
- ☐ Remind call in members that meeting will begin promptly at designated time after making connection. **(Leads or AA)**
- ☐ Call meeting to order at designated time. **(Leads)**
- ☐ If meeting needs to begin later, notify members why meeting is delayed and expected start time. Repeat this every 2-3 minutes until meeting begins. **(Leads)**
- ☐ Take attendance. **(Leads or AA)**
- ☐ Review purpose of the meeting with membership. **(Leads)**
- ☐ Introduce new members and guests. **(Leads)**
- ☐ Remind members to disclose any pertinent potential conflict of interest during any meeting deliberations. **(Leads)**
- ☐ Review previous meeting minutes. **(Membership)**
- ☐ Approve previous meeting minutes. **(Membership)**
- ☐ Work through agenda with intent to stick to allotted time for each agenda item. **(Leads)**
- ☐ If extra time needed for additional discussion of an agenda item, then ask members if it is okay to go past the allotted time. If so, then notify members of what the additional time limit will be. **(Leads)**
- ☐ Review VAT opportunities and decide on top 3. **(Membership)**
- ☐ Form first VAT **(Leads)**
- ☐ Review potential metrics and dashboard with membership. **(Leads)**
- ☐ Decide on which metrics to track. **(Membership)**
- ☐ Request any future agenda items to be forwarded to the Leads and the AA. **(Leads)**
- ☐ Announce next meeting date, time and location. **(Leads or AA)**

☐ Adjourn meeting on time. **(Leads)**

Prior to Third Value Analysis Committee Meeting

☐ Confirm next meeting and teleconference capability. **(AA)**
☐ Confirm availability of audio-visual equipment and support contact information in case of problems. **(AA)**
☐ Send out next meeting notice to membership along with any specific instructions on how to access the meeting via phone or internet. **(AA)**
☐ Schedule meeting with Leads, Contract Specialist, Data Analyst to prepare for first VAT meeting. **(AA)**
☐ Schedule meeting for Leads with Data Analyst to review data. **(AA)**
☐ Meet with Data Analyst to identify opportunities for improvement. **(Leads)**
☐ Pull any data, graphs or charts needed for meeting and forward to Leads. **(Data Analyst)**
☐ Leadership to review any data, graphs, charts or documents for next meeting and forward to AA. **(Leads)**
☐ Leadership to develop a list of 3 – 5 goals for the Value Analysis Committee based on opportunities identified in data. **(Leads)**
☐ Leadership to develop a draft of how the committee will communicate its progress to its stakeholders. This can be in the form of a newsletter, memo or presentation. **(Leads)**
☐ Leadership to develop a communication plan with timeline and list of stakeholders for communication. **(Leads)**
☐ Leadership to develop, confirm and forward final agenda to AA. **(Leads)**
☐ Send out follow-up meeting notice along with final agenda and any documents to membership. **(AA)**
☐ Confirm RSVPs for meeting. **(AA)**
☐ AA to notify Leads of any non-responders to RSVP. **(AA)**

- ☐ Leadership to contact any membership non-responders to meeting notice to confirm RSVP. **(Leads)**
- ☐ Leadership to contact AA with any confirmed RSVP from non-responders. **(Leads)**
- ☐ Arrange refreshments (if any) for next meeting based on RSVP response. **(AA)**
- ☐ Prepare any paper documents needed for membership at the next meeting (agenda and documents). **(AA)**

Third Value Analysis Committee Meeting

- ☐ Arrive 10 – 15 minutes before meeting begins. **(Leads & AA)**
- ☐ Set-up any audio-visual equipment and test. **(Leads or AA)**
- ☐ Set-up any teleconferencing equipment and test. **(Leads or AA)**
- ☐ Call in to any bridge line or web enabled teleconferencing 3 – 5 minutes before meeting begins. **(Leads or AA)**
- ☐ Remind call in members that meeting will begin promptly at designated time after making connection. **(Leads or AA)**
- ☐ Call meeting to order at designated time. **(Leads)**
- ☐ If meeting needs to begin later, notify members why meeting is delayed and expected start time. Repeat this every 2-3 minutes until meeting begins. **(Leads)**
- ☐ Take attendance. **(Leads or AA)**
- ☐ Review purpose of the meeting with membership. **(Leads)**
- ☐ Introduce new members and guests. **(Leads)**
- ☐ Remind members to disclose any pertinent potential conflict of interest during any meeting deliberations. **(Leads)**
- ☐ Review previous meeting minutes. **(Membership)**
- ☐ Approve previous meeting minutes. **(Membership)**
- ☐ Work through agenda with intent to stick to allotted time for each agenda item. **(Leads)**
- ☐ If extra time needed for additional discussion of an agenda item, then ask members if it is okay to go past the allotted

time. If so, then notify members of what the additional time limit will be. **(Leads)**
- ☐ Review metrics and dashboard with membership. **(Data Analyst)**
- ☐ Review timeline and plan for first VAT meeting. **(Leads)**
- ☐ Discuss opportunities for performance improvement with membership. **(Leads)**
- ☐ Propose and choose 1 – 3 opportunities for the committee to use for their first performance improvement project. **(Membership)**
- ☐ Decide on action plans or create teams to develop action plans for performance improvement projects. **(Membership)**
- ☐ Assign Leads and teams to each performance improvement project. **(Leads)**
- ☐ Decide on which metrics to track for performance improvement project(s) and SMART goals. **(Membership)**
- ☐ If needed, then review what "SMART" goals are with membership. **(Leads)**
- ☐ Present communication plan to membership for comments and approval. **(Leads)**
- ☐ Request any future agenda items to be forwarded to the Leads and the AA. **(Leads)**
- ☐ Announce next meeting date, time and location. **(Leads or AA)**
- ☐ Adjourn meeting on time. **(Leads)**

ABOUT THE AUTHOR

Dr. Terrence Loftus is the President of Loftus Health, a healthcare consulting company committed to educating and coaching the next generation of healthcare leadership on how to improve the delivery of healthcare. Prior to this, Dr. Loftus was the Medical Director of Surgical Services & Clinical Resources for Banner Health in Phoenix, Arizona. He obtained a BS in Psychology and an MBA from Arizona State University, and his Medical Degree is from the University of Arizona. He completed a residency in General Surgery at the University of Utah and a Trauma Surgery and Surgical Critical Care Fellowship at the University of Maryland's R Adams Cowley Shock Trauma Center in Baltimore, Maryland. Dr. Loftus is also a graduate of the Advanced Training Program for Executives and Quality Improvement Leaders sponsored by Intermountain Healthcare's Institute for Healthcare Delivery Research. Dr. Loftus has served in various leadership roles including Chief Medical Officer, as well as a Medical Director of a Surgical Intensive Care and a Level 1 Trauma Center. He is board certified in General Surgery and Surgical Critical Care, and is a Fellow in the American College of Surgery. He can be contacted by visiting his website at www.LoftusHealth.com.

www.ingramcontent.com/pod-product-compliance
Lightning Source LLC
Chambersburg PA
CBHW080931170526

45158CB00008B/2251